# Key Concepts in Genetics

# KEY CONCEPTS IN GENETICS

**Abeer Iqbal**

Delve
PUBLISHING
www.delvepublishing.com

**Key Concepts in Genetics**

*Abeer Iqbal*

**Delve Publishing**
224 Shoreacres Road
Burlington, ON L7L 2H2
Canada
www.delvepublishing.com
Email: orders@arclereducation.com



ISBN: 978-1-77469-393-3 (Hardcover)

# ABOUT THE AUTHOR

**Abeer Iqbal (M.S Microbiology and Molecular Genetics)** is a polished writer with strong scientific knowledge and keen interest in biosciences. She has authored books related to her field and have also been engaged in writing medical blogs/content. She has a passion for learning and writing about scientific advancements and her enthusiasm reflects in her work. She is a nature lover and admires the importance and beauty of little things in life.

# TABLE OF CONTENTS

# PREFACE

Genetics is a branch of biology that deals with genes, genetic diversity, and heredity in living things. It focuses on heredity, or how a parent transmits certain genes to their offspring. Offspring inherit their biological parents' genes that express specific traits, for example, one's physical characteristics, natural talents, and genetic disorders.

The majority of living cells contain the 'software' that controls and decides how it is supposed to look, function and perform certain functions at the molecular level. This particular software is scripted as codes on a specific molecule, known as the Deoxyribonucleic Acid (DNA). DNA has the same chemical nature in every life form on earth.

The code defines genetic words which can be translated in to basic biological functions. Such basic genetic codes are known as Genes. All these genes combine to form the Genome (the software of life). Genetic processes can be affected by a combination of factors like an organism's physical and chemical environment and experiences to influence development and behavior.

Genetics studies is research on how individual genes or a genome are involved in health, functionality and diseases. This understanding and knowledge of genetic factors and genetic disorders are important concepts promoting health and preventing disease. Genetic diseases can be categorized into three major groups: single-gene, chromosomal, and multifactorial. Changes in the DNA sequence of single genes, also known as mutations, cause a lot of diseases.

In this book on genetics, we present the basic terms of genetics and key terms related to it like mutations, chromosomes, polymorphisms, and epigenetic terms, etc. The overall aim of the book is for students to develop a complete understanding of basic genetic concepts which they can then apply or build upon as needed.

This book will introduce the readers to the field of genetics and its fundamental terms. It is precisely designed for readers or students with no prior experience related to genetics, and brings up a diversity of fundamental topics. By the end of the book, readers will get a basic idea of genes, genetics, heredity and other key concepts involved.

# A

## A24 Protein (Semi-Histone)

Protein complex ensuing from the combination of ubiquitin to histone h2A. The quantity of h2A histone that is in the form of a24 varies in special cells. Research shows that chromatin which consists of actively transcribed DNA sequences is enriched in a24 protein.

## A Band

One of the bands seen in striated muscle. A bands are exclusive from 1 band which they substitute in striated muscle, the former being composed of myosin fibrils and the latter of actin, troponin and tropomyosin.

## Ab Antigens

Human blood group elements are composed of oligosaccharides and connected to glycolipid in the plasma membrane of erythrocytes, or to soluble glycoproteins in tears, gastric juice and saliva.

Their characteristic IS not known; however, they are detected through the capability of naturally occurring antibodies existing in individuals lacking the corresponding antigen and which probably arise because comparable oligosaccharides exist on cell membranes of enteric bacteria.

## ABO Blood Group System

Human erythrocytes carry both a or b antigen, or both, or neither on their surface, so there are four phenotypes of blood groups: A, B, AB and O. Inheritance relies upon three alleles: lA, 1B, and i. IA and 1B are codominant to i. The alleles segregate to form six genotypes: IAIAi and IAi (group A), 1B 1B and IBi (group B), IAIB (group AB) and ii (group O).

Antigens a and b are sugars derived from H substance via the motion of glycosyl transferases encoded through the genotype. I antibodies, referred to as isohaemagglutinins, A and B show up when antigen is absent from the erythrocyte surface (e.g., an individual of blood group A has anti-B). The ABO system is vital in blood transfusion practice.

0O donors (universal donors) can give red blood cells to any affected person (allother blood group structures are compatible); AB genotype (universal recipients) can receive red blood cells from any donor. The system is also essential in transplant surgical operations and in medicolegal practice for the establishment of paternity.

## Abortive Transduction

Transduction in which donor DNA fragments are stably inherited in the absence of recombination as non-replicating structures inside recipient bacteria. Abortive transduction is used as a check for genetic complementation. On cell division, the exogenote passes to only one of the daughter cells.

The other daughter cell may, however, acquire a cytoplasmically sufficient donor gene product (e.g., an enzyme) to allow persistent expression of the donor trait for numerous cell generations. Abortive transduction is characterized by the presence of small (often microscopic) colonies on media selective for the assayed phenotype.

Only one cell in each of these colonies genuinely contains the exogenote, so the colonies develop non-exponentially. When minute colonies are absent from suitable selective media then no section of the exogenote can compensate for the functional deficiency of the recipient endogenote and the mutations are consequently allelic. The presence of abortive transductions suggests genetic complementation between exogenote and endogenote; this can be interallelic (allelic complementation) or intergenic.

## Abscisic Acid (ABA)

Plant growth hormone, once named abscisin ii, is liable for leaf fall, (abscission), in which old leaves are parted from the parent plant by alterations to the cells in the abscission zone. Auxins are additionally associated with the course of abscission.

## Absolute Plating Efficiency

Percentage of the cells in a preliminary inoculum that produces colonies when plated onto Petri dishes or different containers with a nutritional medium.

## Absorbance

Education in depth of light, or different radiation, passing via an absorbing medium. In the usage of a spectrophotometer, it is portrayed as A=log10(IoI), incident intensity Io and transmitted intensity I. The maximal absorbance of nucleic acids is around 260 nm and that of protein at 280 nm.

## Absorption Spectrum

The spectrum obtained when ultraviolet or infrared light is passed through a substance. Light sources such as hydrogen or tungsten lights are utilized. The light passes through channels or gratings to guarantee that it is monochromatic (i.e., of a single fixed wavelength at one time).

Graphs of relative retention of the different frequencies, as estimated by a light-sensitive cell, can be utilized to show the presence of biologically significant substances. Colored compounds (e.g., hemoglobin and chlorophyll) have specific absorption spectra.

For instance, chlorophyll seems green on account of its reflectance of green light and absorbs light in the blue and red parts of the visible spectrum. Proteins retain most firmly at around 280 nm and nucleic acids at 260 nm.

Instruments that are utilized to gauge the absorbance of visible light are regularly called colorimeters, though those estimating a greater range scope of frequencies are separated using spectrophotometers.

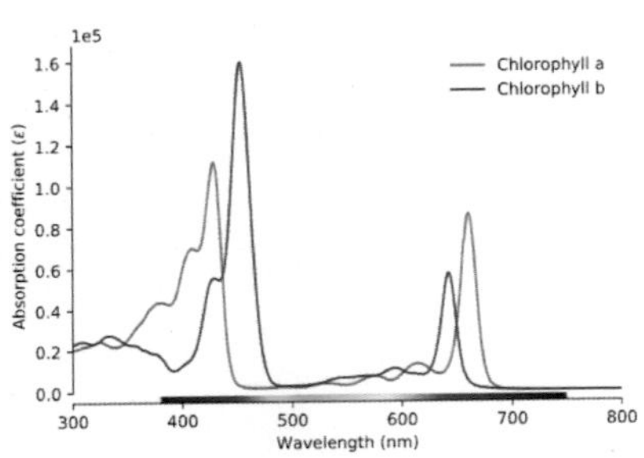


**Figure 1.** Absorption Spectrum Of Chlorophyll A And B.

*Source : Image By Wikimedia Commons*

## Absorptive Endocytosis

Take-up of material from the encompassing medium by the means of dynamic endocytosis through the plasma layer of the cell.

## Acceptor Splicing Site

The splicing site  at the 3' right-hand end of an intron.

## Acceptor Stem

Direct augmentation of an exchange RNA to which a particular amino acid is connected. For the most part, the expansion is double-stranded, however, it ends at its 3' - end in a triplet of non-paired bases, which are CCA.

## Accessory Cell (Subsidiary Cell)

Plant cell in the epidermis that encompasses or is adjacent to a guard cell. Accessory cells help the guard cells during stomatal opening and shutting.

## Accessory Nuclei

Small nuclei seen in the oocytes of some hymenopteran insects. They are obtained from the significant oocyte nucleus.

## Acellular

Demonstrating a tissue or organism that comprises a mass of protoplasm which is not generally partitioned into cells. Multinucleate hyphae of certain mold and the plasmodium of the slime mold, Physarum are examples of a cellular states.

## Acentric

A chromosome or chromatid that lacks the center part known as the centromere. Portions of chromosomes that might broke off and lack centromeric regions could also be characterized as acentric.

## Acetabularia

A genus of comparatively large unicellular green algae. Acetabularia cells have been widely used in hereditary experiments. In this study, nucleated basal fragments of one species are combined with anucleate fragments of another species to shed light on the genetic roles of the nucleus and cytoplasm.

## Aceto-Carmine

Dye consisting of 5 percent carmine solution in acetic acid; previously used as a chromosome stain, however now mostly changed with the aid of aceto-orcein.

## Aceto-Orcein

Dye consists of a solution of orcein in acetic acid. Squash preparations of chromosomes are regularly stained with aceto-orcein.

## Acetylcholine

Compound that is accountable for the transmission of the nerve impulse throughout the synapse between adjacent nerve cells or across the neuromuscular junction. Acetylcholine molecules combine with specific receptors in the surface membrane of the postsynaptic cell.

Initiating a new action potential, the molecules are unexpectedly degraded by means of the action of cholinesterase making sure that the impulse is transient. Acetylcholine is the acetyl ester of the natural base choline and on breakdown yields acetate and choline.

## Acetylmuramic Acid

Disaccharide is located in linear polymers in peptidoglycan and takes place in the periplasm between the internal and outer membranes of gram-negative bacteria.

## Achiasmate

The absence of chiasmata during meiosis. Crossing-over is restricted to one sex in some species, such as Drosophila, and in these cases, the heterogametic sex has chiasmata meiosis.

## Achondroplasia

Dwarfism is inherited as an autosomal dominant gene in humans. It affects about one in every 10,000 individuals. It is distinguished by long bone shortening caused by abnormal endochondral ossification, but affected individuals are both intellectually normal and sexually fertile.

## A Chromosome

Normal chromosomes of any eukaryotic cell, as opposed to supernumerary chromosomes or B chromosomes.

## Acid Fuchsin

Dye has an acidic organic group that determines staining properties.

## Acid Hydrolases

Enzymes that catalyze hydrolytic reactions, are active within lysosomes, and are most active at about pH.5.

## Acidic Amino Acids

At a neutral pH, amino acids with a net negative charge. Glutamic acid and aspartic acid are two examples of amino acids with negatively charged side chains at normal cellular pH. levels.

## Acidic Dye

Dye (e.g., acid fuchsin) mainly stains positively charged material asthe dye includes a net negative charge.

## Acidic Protein

Protein with a net negative charge at neutral pH. Due to a high concentration of acidic amino acids. Most proteins in eukaryotic cells are acidic, with the fundamental proteins - the histones - being notable exceptions. Some acidic proteins are thought to play critical roles as gene-regulatory molecules.

## Acid Phosphatase

The enzyme that splits phosphate groups from a variety of resources and is most active at low pH. It is a necessary component of lysosomes and was initially used as an assay to identify cell fractions rich in lysosomes. There is evidence that acid phosphatase is packaged in the cell's Golgi apparatus.

## Acinar Cell

An exocrine pancreatic cell that produces digestive enzymes and zymogens such as amylase, lipase, trypsinogen, chymotrypsinogen, DNase, and RNase.

Acinar cells have revealed details about cellular secretion processes: for example, pulses of radiolabeled amino acids are delivered to the cells, and the amino acids' subsequent fate is tracked using autoradiography.

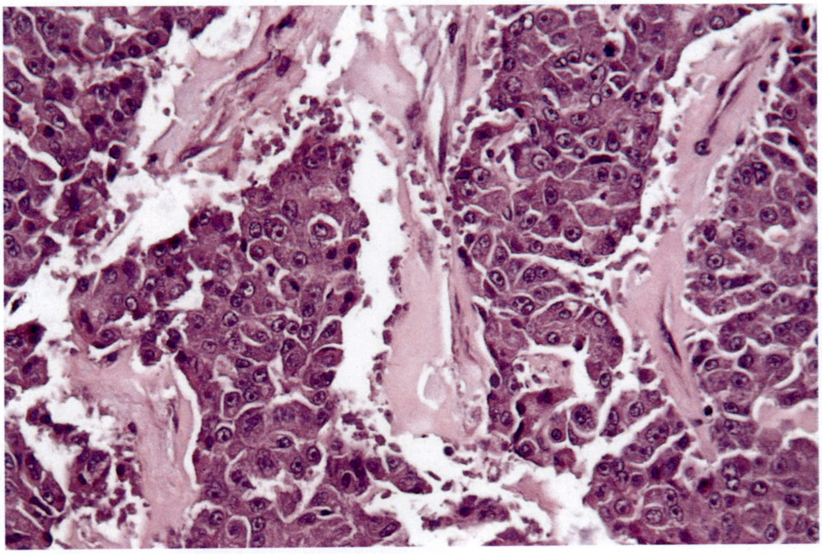

**Figure 2.** Acinar Cell – Carcinoma Of The Pancreas.

*Source : Image By Wikimedia Commons*

## Acinus

Fundamental secretory unit of an exocrine organ. It has a saccular or tubular shape, the wall of which is ordinarily cuboidal or columnar epithelial cells.

The secretory product is set free from the apical cell surfaces into the encased lumen from which it channels into a group pipe. In a compound exocrine gland, the acini might look like a bunch of grapes (acinus = grape) and channel into a complicated, stretched ductal system.

## Acrasiales

Group of soil organisms, perhaps related to fungi, and frequently described as the cellular slime molds. They encompass such organisms as Dictyostelium. Although the Acrasiales are regularly positioned collectively with the myxomycetes in the massive umbrella of slime molds - Myxomycophyta - the two groups are probably now not intently related.

The slime molds, in contrast to the myxomycetes, have a multicellular structure for most of their life cycle recognized as a pseudoplasmodium. Acellular slime molds such as *Physarum* possess an authentic plasmodium that lacks cell membranes and is as a consequence genuinely a very giant coenocytic structure.

## Acrasin

The chemotactic molecule that is liable for initiating conglomeration of scattered amoebae of cellular ooze molds Acrasiales to frame the colonial pseudoplasmodium additionally named a slug or grex. Acrasin has now been distinguished as a cyclic amp.

## Acridines

Class of natural molecules with two rings that render them of top-notch natural significance. They ARE mutagens, causing augmentations or deletions of bases in DNA, particularly in plasmids, bacteriophages, and mitochondria. Numerous acridines have the property of intercalating between connecting base pairs of DNAs thus causing loosening up of the helix.

Many are moreover fluorescent in ultraviolet mild and form useful fluorescent stains for such cell organelles as nuclei and nucleoli. Examples of acridines are acridine orange, acriflavine and proflavine. Some have been used as cytotoxic medications in many cancer chemotherapy and bacterial disease.

## Acridine Orange

One of the acridines: causes modifications such as accumulating in body tissues as well as other transformations such as nucleotide deletions or additions.

It is therefore useful as a fluorescent stain in living organisms since acridine orange binding will give DNA-containing structures a green fluorescence and RNA-containing structures an orange fluorescence under UV light. It also reduces the replication of some bacterial plasmids which can be used to remove f portions from f+ *E. coli*.

## Acriflavine

As an antiseptic, the yellow substance is used. It is a compound composed of 2,8-diamino-10-methylacridinium chloride and 2,8-diaminoacridine dihydrochloride. Acriflavine is also a potent mutagen, and it is particularly useful in inducing the petite mutant in yeast.

## Acrocentric

Describes a chromosome in which the centromere is significantly off center. If the centromere is actually located at one end of the chromosome, the term telocentric is used.

## Acrosome

A spermatozoon's head area is an organelle located at the apex of the sperm head that is secreted by the Golgi apparatus of the developing sperm and is responsible for the digestion of the egg cell membrane.

## Acrosome Reaction

The first reaction of an egg cell is when it comes into contact with a spermatozoon. The sperm's acrosome vesicle, a package of hydrolytic enzymes, is launched by exocytosis, digesting a path for the sperm through the egg's zona pellucida and allowing it to fuse with the egg cell plasma membrane. Invertebrate sperm, the reaction also produces specific protein molecules that aid in the binding of the sperm to the egg coat.

## ACTH (Adrenocorticotrophic Hormone, Corticotrophin)

Polypeptide molecules are discharged during stress by the anterior region of the pituitary organ of warm-blooded organisms. ACTH's belongings start the synthesis of corticosteroids in the cerebrum and of melanin in pigment cells.

## Actin

The primary protein component of muscle and other contractile structures (e.g., cilia and flagella). Actin molecules are filamentous proteins made up of two strings of globular units (G actin) that are wound around one another to form a rope-like structure (F actin).

## Actin Filament

Actin's filamentous structure is made up of two strands of globular molecules, each about four nanometers in diameter, twisted into a helix with 13.5 molecules per turn. These filaments have been identified as the primary contributors to thin filaments in striated muscle.

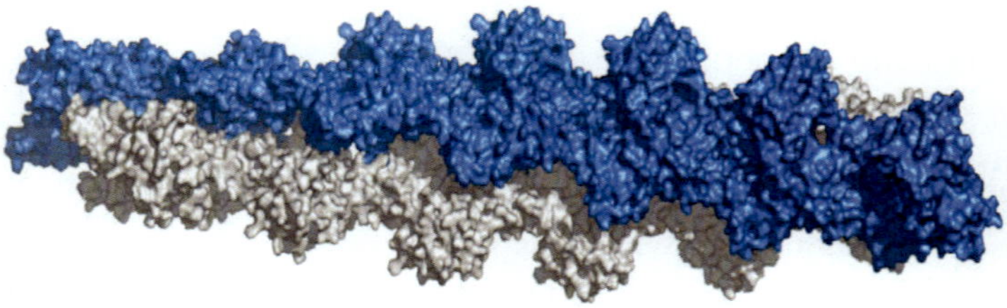

**Figure 3.** Molecular Structure Of An Actin Filament.

*Source : Image By Wikimedia Commons*

## Actin Genes

Actin-coding genes are an important gene family in animal cells. In mammals, up to 20 distinct coding sequences have been scored, but not all are necessarily transcribed.

Although some actin sequences are linked, others are widely dispersed throughout the genome. In mammalian cardiac and skeletal muscle, distinct actin genes are activated. In the actin gene family, at least one pseudogene has been discovered.

## Actinomycin D

Antibiotics inhibit RNA synthesis by intercalating between the bases of DNA, particularly in cytosine and guanine-rich regions. The drug has a particularly strong effect on ribosomal RNA synthesis.

## Action Spectrum

The spectrum wherein the size of a molecule is estimated in contrast with the frequency of light (typically ultraviolet to near-infrared). Exhibiting a functional connection between the presence of a light-absorbing pigment and the recorded exercise is utilized.

## Activator Element (AC)

Transposable elements have been found in the maize plant by Dr. Barbara Mcclintock. Its product is in a cooperative fashion with those of various transposable elements in maize. Specifically, the transposable element will possibly bring about breakage in chromosomes when an activator is additionally present.

## Activator Molecule

A molecule that especially activates or induces a biochemical reaction; is most often utilized for molecules that act as gene regulators.

## Active Gene

Gene sequence that is by and large effectively interpreted. Since the paces of record fluctuate, the movement of genes is not a ·au-or-nothing' peculiarity. A few genes, like those coding for tRNA, are interpreted at a high rate over the lifetime of the cell and are supposed to be constitutive. Others.

ike those coding for histones, are expressed in all cells however are active just at specific times during the cell cycle, and some (e.g. .. Globin gene), are active possibly in blood-forming tissue during times when hemoglobin formation is required. Active genes have specific biochemical attributes, some of which might be somewhat liable for the course of initiation, others might follow from the reality of actuation.

he attributes of most active genes are: ( 1) that they are in an open configuration in chromatin and along these lines all the more effortlessly processed by enzymes, for example, DNAase I; (2) that they have specific acetylated histones related with them; (3) that they are poorly methylated; (4) a few active genes are additionally connected with histone that is enriched for a24 protein.

## Active Immunity

Development of an immune response with the aid of the host's immune device following injection. Or after exposure to, an antigen, and often ensuing in long-term safety must the antigen be a pathogen or some inactivated structure of it.

## Active Site

art of an enzyme that specially binds the substrate(s) and catalyzes product formation. The active site consists of two, regularly overlaying, regions: the binding site, which recognizes and binds the substrates(s); the catalytic site, which catalyzes the response as soon as binding has occurred. The amino acids in each of these regions are no longer necessarily adjoining to the linear polypeptide however are added collectively with the aid of the folding of the protein molecule.

The active site, whether or not going on the surface of the enzyme or buried in a cleft, occupies only about 5 percentof the whole surface of the enzyme molecule. The initial binding of the substrate includes the formation of non-covalent bonds (e.g., hydrogen bonds, electrostatic bonds, hydrophobic interactions, van der Waal's forces with chemical molecules at the active site.

These molecules may additionally cross-linkbarely in order to accommodate the substrate (the brought about fit theory). So, the active website residues in a flexible vicinity of the protein no longer normally involve inflexible structures. During catalysis, covalent bonds may be formed, and then broken, as a section of the response mechanism.

Catalysis generally includes one or greater of the following: most suitable spatial alignment of the substrate(s) on the enzyme surface: distortion of bond angles and stretching of bond lengths; transfer of protons or electrons (acid/base catalysis).

The chemical molecules in catalysis include the chemically reactive sidechains of the amino acids histidine ,lysine, arginine,,, serine, threonine, tyrosine, cysteine, glutamic acid and aspartic acid. And for some enzymes a coenzyme or cofactor.

## Active Transport

ransport of molecules against a concentration gradient across a biological membrane. This suggests the development of molecules from areas of low to areas of high concentration, the converse of direct diffusion. Active transport ordinarily includes transporter proteins, which tie the particle to be transported and move with it through the layer.

Energy is additionally required and is provided by a tp. Additionally suggests that a means of active transport is directional and not really reversible. Significant examples of active transport are the transport of sodium from neurons to start a resting potential and the import of amino acids into cells to allow protein synthesis.

## Actomyosin

A protein that is an actin-myosin conjugate. The conjugate is formed on a temporary basis during striated muscle contraction.

## Adaptation

Change in a nervous receptor with regular stimulation, such that fewer impulses per unit time are evoked by means of an exterior stimulus. Adaptation may also result from modifications in the membranes of receptor cells, or it can also comply with adjustments in accessory structures associated with a sense organ. The sharp decline in sensitivity to a specific odor in the human is an example of adaptation.

## Adaptive Enzyme

The enzyme is synthesized solely in the presence of an inducer, commonly a substrate molecule. Production of such enzymes involves an adaptation of the cell to an alternate in the exterior or interior environment.

# B

## Backbone

Linear, supporting the lower back of a polymer molecule from which the sidechains project. The backbone of DNA consists of two antiparallel strands of alternating sugar and phosphate groups to which the bases attach and from which they achieve support. In the case of DNA, the formation of hydrogen bonds between the bases within the helix gives further strength to the structure.

## Backcross

Cross-breeding occurs when an individual from the F 1 generation marries a parent or an individual with the same genotype as the parent.

## Background Radiation

Radiation from natural sources, such as cosmic radiation and granitic rocks, is in contrast to radiation from radioisotopes involved in experimental situations. Fallout from atomic explosion might also contribute extensively to background radiation, as will accidentally leaks from atomic electricity stations.

## Back Mutation

The mutation that reverses a previous mutation, frequently however no longer continually restores the wild-type genotype. Thus, if a wild-type Drosophila population were screened, a mutant with vestigial wings might be discovered. If a population of such flies has been produced, some again mutations might be identified in which the wild-type genotype used to be restored.

## Bacteria

Simple prokaryotes; are organisms occuring in the soil, water and air, and as symbionts, parasites and pathogens of animals, plants and different microorganisms. They represent the amplest cell type on earth. Although unicellular, the cells may also be related as chains, clusters or mycelia (as with positive actinomycetes).

The linear dimensions of microorganisms are approximately 1-10 p.m. (some, e.g., chlamydia, are less than 1 p.m., whereas others, e.g., spirochaetes, are up to five hundred p.m.). The fundamental shapes of microorganisms are the bacillus (rod), coccus (sphere) and spirillum (spiral). Bacteria often possess a tough protective cell wall, below which a membrane surrounds a single cytoplasmic compartment.

Bacteria reproduce asexually by binary fission, regularly with high-quality rapidity (generation examples of less than 20 minutes have been recorded). Morphological differentiation is generally absent, although some species (e.g., bacillus) form spores.

Most microorganisms are heterotrophs, but a few are autotrophs utilizing either the strength from chemical oxidations (chemolithotrophs) as with *Nitrobacter* and *Thiobacillus* or radiant power (photoautotrophs) as with *Chlorobium*. Metabolism may additionally be by means of fermentation or by way of aerobic or anaerobic respiration.

## Bacterial Conjugation (Bacterial Mating)

Directional transfer of genetic material (DNA) through cell-cell contact between the donor (male) and recipient (female) bacteria. Conjugation is one mechanism by which plasmids unfold through bacterial populations and is additionally the basis of the classical approach to the genetic evaluation of bacteria.

First determined for *E. coli* in 1946, conjugation has been found for different gram-negative bacteria which include species from *Azotobacter, Klebsiella, Pseudomonas, Salmonella, Shigella and Vibrio*, and between species from some genera (e.g., *E. coli* x *Salmonella, Salmonella* x *Shigella*).

Donor DNA is conferred by means of an intracellular plasmid sex factor. The plasmid is of the conjugative (self-transmissible) kind (e.g., the f factor, resistance aspect or sure large colicin factors). In f+ donor cells, the f factor is present as one or two extrachromosomal copies. Recipient (f-) cells lack the f plasmid.

Donor-recipient mating pairs are fashioned following random collisions quickly after mixing cultures of f+ and f- bacteria. The cells are then related via the f pilus, a proteinaceous appendage of the donor. The tip of the pilus interacts with a specific receptor protein on the wall of the F-cell (a main outer membrane protein).

The role of the pilus can also be to convey the cells together with the aid of pilus retraction through depolymerization. During conjugation, F- bacteria gain the F-factor and are therefore converted to F+. Transfer of the F plasmid is dependent on DNA synthesis. One strand of the circular, double-stranded Fplasmid is cut at a particular site orit (origin of transfer) by means of a plasmid coded endonuclease (product of the tray and traz genes).

Unidirectional DNA synthesis, the usage of a membrane-bound DNA polymerase, is initiated at this site, the unbroken round strand appearing as a template. The newly synthesized strand displaces the reduced strand, which then enters the recipient (5'-end first). The transferred donor strand is replicated in the recipient telecell to form a double-stranded f factor.

All the f- cells in a culture may be transformed to f+ donors in this fashion, whereas the original donor cells hold their donor character. In about one in a hundred zero f+ cells the f component is stably built-in into the chromosome, thereby losing its autonomy and behaving as a single replicon with the chromosome.

hese bacteria do not often confer the donor properties on f- recipients, however, hfr x f- crosses supply an excessive frequency of recombination of donor markers. In a given hfr pressure integration of the F element has occurred, with a specific orientation, at one of several precise sites scattered around the chromosomes.

These sites show some sequence homology with f-DNA. The integration technique is reversible main to the reconstruction of the plasmid F factor. Mobilization of the chromosome is similar to f transfer. During conjugation, a cut is made in one DNA strand at orit within the f factor, whereupon the 5'-end unwinds and enters the recipient.

These consequences in the immediate entry of phase of the F element are accompanied by means of the oriented switch of chromosomal genes, proximal to the main end, at a uniform price of about 1 percent of the chromosome length per minute. The donor remains potential because DNA synthesis replaces the transferred strand.

DNA synthesis in the recipient additionally converts the transferred single strand into double-stranded DNA. The order in which the donor genes are transferred is decided via the position and orientation within the chromosome. Only when the whole size of the chromosome is transferred (about 100 minutes) is the whole fFissue transmitted and HfrDNA conferred on the recipient.

Complete chromosome transfer occurs very rarely (from about one in 10 zero donor cells); strand breakage usually terminates switch prematurely. The presence of donor DNA inside the recipient creates a merozygote, which is diploid for the genes carried on the fragment.

The fragment is not an independent replicon and is lost at cell division except it turns into integrated into the recipient chromosome with the aid of crossing over at two or more sites to create a stable recombinant.

F' elements are excised from the chromosomes of a hfr stress and consist of the f factor together with adjacent areas of the donor chromosome (up to one-third of the chromosome may also be attached). F' formation takes place in main donors, which thereby endure chromosome deletion.

During x F- conjugation (f duction), F' is transferred in a manner analogous to F transfer; the recipient becomes a secondary f' donor and is diploid for the chromosomal genes carried by using the f' factor. F' can be reversibly built-in into the chromosome within the diploid place so that cultures of intermediate donor lines are about 90 percent of and 10 percent of bacteria. F' factors are used easily to map gene loci on the bacterial genome and for complementation analysis of genes.

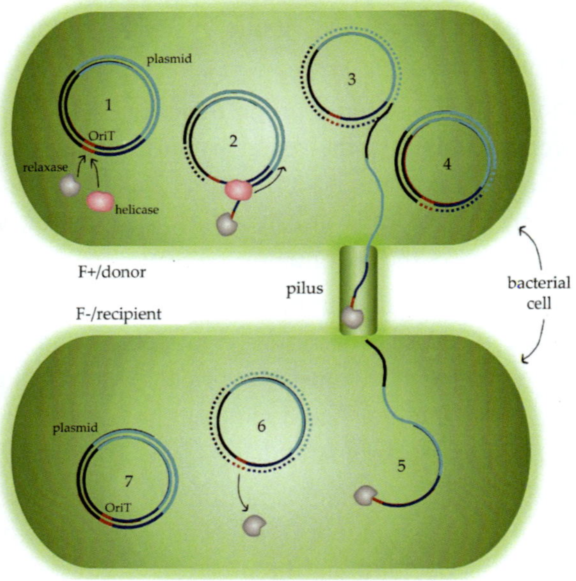


**Figure 4.** Image Showing Bacterial Conjugation.

*Source : Image By Wikimedia Commons*

## Bacterial Lawn

Persistent, uniform layer of microscopic organisms developing on the surface of a strong medium such as an agar plate. Lawns are arranged by flooding the agar plate with a suspension of microbes, taken after by a brooding period to permit bacterial development. Restraint of development inside a lawn is a sign of the affectability of the life form to agents connected to the lawn (e.g., anti-microbials, bacteriophages).

## Bacterial Slant (Bacterial Slope)

Slants are utilized for putting away stock cultures of microbes, yeasts and other microorganisms within the research facility. Microorganisms are developed on the surface of an appropriate nutrient medium set at 45° from even in a little, airproof vial. The strategy is most reasonable for aerobes, but the slants of anaerobes can be secured with sterile fluid paraffin.

## Bacterial Stab

Stabs are used for storing stock cultures of microorganisms in the research facility. The microorganisms are inoculated into a smooth medium (e.g., 0.6 percent wt./vol nutrient agar) contained inside a small, airtight vial with the usage of a straight wire. The method is appropriate for anaerobes and facultative anaerobes.

## Bacterial Transposon

Any one of a range of distinct DNA sequences determined in some species of bacterial cells that are characterized through their mobility in the genome, considering they can go from one site to any other inside the circular chromosome. They owe their mobility to the existence of short insertion elements at each end of the transposon, and they might also elevate bacterial coding sequences within the transposon itself.

Transposons are detected in microorganisms by the movement of a coding sequence within a transposon (as demonstrated by mapping the gene sequence using an interrupted mating experiment) or by the altered expression or mutation rate of other bacterial genes adjacent to the site of insertion of a transferable sequence.

## Bactericidal

A chemical or physical agent that has an irreversible lethal effect on at least some types of bacteria.

## Bacteriochlorophyll

Some autotrophic bacteria contain a chlorophyll-like compound. It has a similar structure to chlorophyll.

## Bacteriocin

A protein secreted by certain bacteria that is deadly (i.e., Bactericidal) for different bacteria. Bacteriocins are commonly named after the producing pressure (e.g., coucins are produced by using *E.coli* and other enteric bacteria). They bind originally to cell surface receptors on the bacterial outer membrane.

They have potent killing action; frequently one molecule is adequate to render a target bacterium non-viable. Many bacteriocins are enzymes acting, for example, on the cell wall mucopeptide (e.g., pesticin from *Yersinia pestis*), or on cell nucleic acids or on membrane factors (e.g., megasin from *Bacillus megaterium* is a phospholipase).

Other bacteriocins spoil the integrity of cell membranes. Colicins a, el, ia, ib and k form ion-permeable channels to target the membrane proton motive pressure. Pyocin rl (produced by *Pseudomonas aeruginosa*), which resembles a contractile bacteriophage tail and includes over 20 protein subunits, additionally damages the energy-transducing membrane.

Most bacteriocins are encoded via plasmids, some of these being conjugative. Bacteriocin synthesis is ordinarily repressed; they are formed due to the fact a cell now and again escapes from repression (this causes the telecell to lyse).

The remaining bacteriocinogenic microorganism in the culture is immune to their very own bacteriocin, some may have an immunity protein that irreversibly binds to and inactivates the bacteriocin. Bacteriocins may additionally confer a selective advantage on the producing stress in its herbal environment by inhibiting the increase of closely associated species with similar nutrient requirements.

## Bacteriolysis

Rupture of bacterial cells. Bacteriolysis can be caused by way of the motion of enzymes (e.g., lysozyme), complement, bacteriophage multiplication or chemical agents (e.g., penicillins, some disinfectants) or mechanical ability (e.g., ultrasonically).

## Bacteriophage (Phage)

Virus infecting and replicating within a bacterial host cell. Bacteriophage research contributed to an understanding of gene regulation, recombination, DNA replication and self-assembly of macromolecules. Phages consist of an outer protein coat (the capsid) surrounding the viral nucleic acid. Other aspects (e.g., lipids, polyamines, ATP) occur in some phages.

Capsids of the simplest phages (e.g., MS2) include one predominant protein type, however over 50 specific protein types are current in complex phages (e.g., T4). The small isometric phages (e.g., 0x174, ms2) have polyhedral (icosahedral) capsids about 25 nm in diameter (often with floor spikes or knobs), whereas the large polyhedral phages (e.g., T phages) are 50-100 nm across.

Many of the large phages have tubular tails with helical symmetry that differ noticeably in dimensions (up to a hundred and fifty nm long) and complexity. The tail may additionally be encased in a contractile sheath and raise long tail fibers. The capsids of filamentous phages (e.g., Ml3) have helical symmetry and are about 1000 x 10 nm in size.

Phage nucleic acid may additionally be DNA or RNA and is linear (e.g., T phages and lambda, double-stranded DNA; ms2, single-stranded RNA) or a circular loop (e.g., M13 and 0x174, single-stranded DNA). The complexity of the phage virion is mirrored in the dimension of the genome (3569 nucleotides code three genes of MS2, 165 kilobases code about 150 genes of T4).

A wide variety of microorganisms are susceptible to phage infection, and given stress may additionally be attacked by using numerous phages. A phage infects solely particular species or strains of microorganisms (the host range). Phages attacking *E.coli* has=ve been most fully investigated.

The phage adsorbs onto the host cell through the attachment of phage receptors (e.g., capsid spikes, tail fibers) to specific receptors on the bacterial surface (commonly a component of the cell wall, however on occasion some other external appendage such as the flagellum or the f pilus).

Changes in the shape of the bacterial receptor following a mutation can lead to phage resistance; this can be overcome by a compensating host-range mutation in the phage. Adsorption is followed by means of passage of the phage nucleic acid into the host cell (for some filamentous and RNA phages part or all of the capsid may additionally also enter the cell).

Entry of viral DNA is assisted in T-even phages by using contraction of the tail sheath causing the tail to penetrate the cell wall. Once delivered into the host cell, phage DNA may additionally be degraded through host restriction endonucleases or, if included via the recreation of host modification enzymes, may also provoke intracellular phage development.

Phage t7 encodes an inhibitor of the host restriction enzyme. The RNA polymerase of the host bacterium is used to transcribe the DNA of simple phages and produce phage messenger RNA (mRNA). For many massive DNA phages (e.g., T phages), where transcription of the phage genes occurs in an orderly sequence of early gene expression followed with the aid of late gene expression, the host RNA polymerase can also undergo modification by phage-encoded proteins.

Some phages specify their very own RNA polymerase. Phage genes are regularly clustered on the genome according to function and order of expression. Transcription can be excluded from one strand of DNA, or from each strand. For RNA phages, the RNA from the virus serves as mRNA.

Translation of phage mRNA into phage proteins utilizes the ribosomes and, most often, the transfer RNA (tRNA) of the host cell. Small phages rely almost absolutely on the machinery of the host to replicate their DNA. Phages of intermediate measurement (e.g., a. Encode a few of their very own enzymes for replication), however, rely notably on host enzymes.

Large phages (e.g., T4) have many genes (30 or more) involved with the synthesis of DNA precursors and with proteins that make up a replication apparatus. RNA phages must of necessity utilize a phage-encoded replicase to produce RNA genomes from an RNA template.

Replication of single-stranded phage DNA or RNA proceeds through the formation of a double-stranded intermediate: the replicative form (RF). RF consists of the specific viral plus-strand and a complimentary, newly synthesized, minus strand.

For DNA phages, rolling circle replication of the round parental rf generates progeny rf molecules which, in turn, produce many copies of circular, single-stranded (+) genomes (synthesis on the (+) strand template being suppressed at this stage by means of the accumulation of particular phage-encoded proteins).

Production of double-stranded phage genomes is by semi-conservative replication and may additionally proceed via rolling circle intermediates Or linear molecules, frequently with the formation of concatemers. Maturation of the easiest phages occurs through a technique of self-assembly whereby phage protein and nucleic acid combination spontaneously produce complete virions. With the extra complicated phages, manufacturing of the virion is not via self-assembly alone but takes place through a collection of obligatory sequential steps. Some proteins are altered during the interaction through cleavage and others have a transient function as 'scaffolding proteins' and are now not retained in the completed virion.

Concatemeric DNA enters preformed empty heads, and when the head is filled the extra DNA is trimmed off via an endonuclease of some phages (e.g., a.). Itrecognizes precise sequences on the genome which therefore decide the length of DNA packaged. Intracellular phage improvement culminates in the release of mature virions from host bacteria.

Filamentous phages are extruded from host cells (which remain viable); most different phages cause cell lysis by producing proteins (e. G., lysozyme) that break the cell wall. The average number of phage molecules released on cell lysis (burst size) levels is from under 20 to over one thousand per cell.

The other dimension relies upon the phage-host device and growth conditions. Temperate phages, unlike virulent phages, try not to continuously get the lytic reaction, yet may go into a steady and quiet relationship with host organisms.

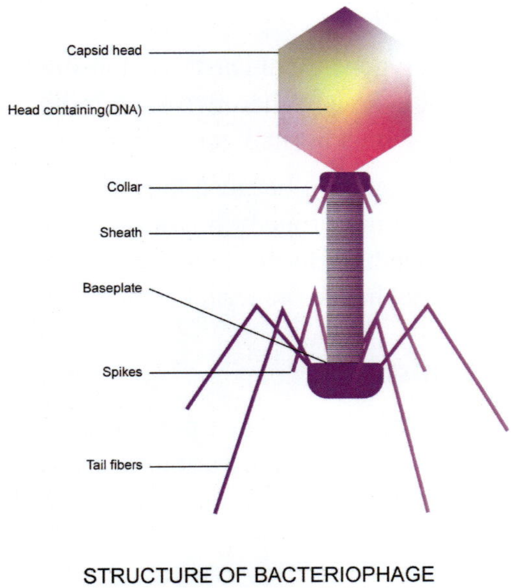


**Figure 5.** Structure Of Bacteriophage.

*Source : Image By Wikimedia Commons*

## Bacteriophage Packaging

Assembly of novel genes into phage genomic DNA, typically of lambda phage, followed by incorporation of these recombinant molecules into phage heads, followed by growth of such phage in host *E. coli* bacterial cells and resulting in the formation of phage plaques.

## Bacteriorhodopsin

Bacterial protein (MR = 25 000) is situated in a specific region of the cell membrane of halobacteria, known as the purple membrane. The protein changes over light into a transmembrane protein gradient that is used to synthesize ATP.

## Bal3l Exonuclease

Exonuclease from the bacterium *Brevibacterium albidum*, which can digest DNA at both the 5' and 3' ends and can be used to insert deletions into cloned DNA molecules.

## Balanced Lethal System

Organisms that are heterozygous for at least two non-linked, recessive genes that are lethal once homozygous. When two such individuals interbreed, a dihybrid cross is formed, with a normal genotype ratio of 9:3:3:1.

The first nine progeny survive because they are heterozygous for both lethal genes, but the remaining progeny die because they are homozygous for at least one of the two lethal. Because the lethal often go undetected, the strain appears to breed true, hence the term balanced lethal.

## Balanced Polymorphism

A genetic polymorphism that persists in a population. The explanation for the balance (i.e., the persistence of the various morphs in their respective ratios) is usually that heterozygotes for the varying alleles have a higher adaptive value than either homozygote.

## Balanced Stock

Heterozygous genetic stock or population that can be kept going for many generations without the use of artificial selection. Such stocks frequently have balanced lethal genes or some other genetic balancing mechanism, such as homozygous female sterility.

## Balbiani Ring

Certain giant puffs of chironomus polytene chromosomes; named after their discovery by Balbiani in 1881. They are highly active genes and sites of intense RNA synthesis, with the genes involved coding for proteins. Secreted in the salivary secretion.

## Band III Protein

Protein found in the erythrocyte plasma membrane. It is a transmembrane protein that is thought to be responsible for the exchange of gaseous oxygen and carbon dioxide across the membrane. The protein is named after its electrophoretic position in relation to other membrane proteins.

# C

## C57 Mouse Cells

Cells from inbred, genetically defined c57 strain mice. C57bl mice were created by a mating in 1921, which also produced c57br and c57l strains. C57bl mice are the most commonly used inbred mice, with the c57bl/6 subline accounting for 14% of all inbred strains used. For *in vitro* research, a number of standard cell lines derived from c57 mouse neoplasms are available.

## Caenorhabditis elegans

A small (1 mm long) free-living nematode that has been widely used in genetics and cell lineage research.

## Cesium Chloride Gradient Centrifugation

Technique of Separating macromolecules, particularly nucleic acids, based on their buoyant densities. An initially homogeneous solution of cesium chloride and nucleic acid is rapidly centrifuged to achieve an equilibrium between diffusion and centrifugal force.

As cesium ions concentrate distal to the axis of rotation, a density gradient forms in the centrifuge tube, whereas DNA and RNA molecules accumulate and band at characteristic points from which they can be recovered.

## Caffeine

It is an alkaloid found in tea and coffee that acts as a stimulant. It is a purine analog that is an active mutagen at high concentrations, particularly on prokaryotic cells, due to its inhibition of specific DNA repair enzymes.

## Calcitonin (Thyrocalcitonin)

One of the hormones that regulates calcium levels in the blood, along with parathyroid hormone (PTH) and the vitamin D derivative 1,25-dihydroxycholecalciferol (1,25-dhcc). Calcitonin, which is produced by the thyroid's c cells, produces hypocalcaemia by inhibiting calcium release from bone.

## Calcium Ion Pump

Chemical pump that is responsible for: (1) keeping the calcium concentration of the cytosol lower than the extracellular concentration by a ratio of approximately 10-7 m inside to 10-3 m outside; as well as (2) pumping calcium ions from the cytosol through the sarcoplasmic reticulum of muscle cells to balance calcium loss due to nerve impulse depolarization of the muscle cell membranes.

## Callus

(1) An unorganized mass of parenchymatous cells that forms at the site of wounds in plant tissue. When cultured under the right conditions, excised pieces of plant tissue and, in some cases, isolated cells can be induced to form callus *in vitro*. This callus can often be kept in culture for a long time if parts of it are transferred or subcultured to fresh nutrient medium on a regular basis. When a callus is transferred to an agitated liquid medium, the tissue disperses and a plant cell suspension culture forms. Some plant callus cultures can be induced to form shoots and roots through organized development.

(2) the formation of new tissue around and in the repair of a fracture, between the ends of bone fragments. It forms as a result of the

proliferation of osteogenic cells, which generate both cancellous bone and cartilage. This cartilage is eventually replaced by cancellous bone, which will then be remodeled into compact bone.

## Calmodulin

Calcium-binding protein found in all animal and plant cells that regulates calcium-dependent enzymes such as adenyl cyclase. A typical animal cell may contain 107 molecules of this protein (1 percent of total cell protein). It is a single polypeptide with 148 amino acids that binds to calcium ions allosterically.

## Cambium

Lateral meristematic tissue of vascular plants, which is found as tissue in long cylinders running parallel to the plant's long axis. It is in charge of supplying cells involved in secondary growth, either in the vascular bundles (vascular cambium) or the outer periderm.

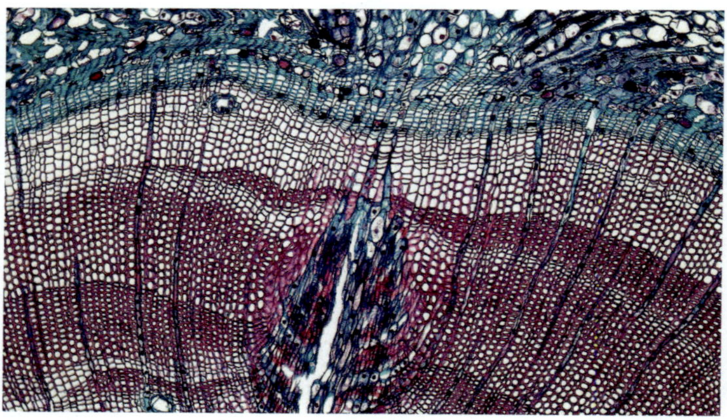

**Figure 6.** Gymnosperm Stem: Vascular Cambium In Five Year Pinus.

*Source : Image By Flickr*

## Campbell Model

A. Campbell proposed a mechanism to account for lambda phage integration into the *E. coli* genome without bacterial DNA loss (as would be expected in conventional recombination). The mechanism, which has been demonstrated to be effective in nature, involves the initial circularization of the linear lambda phage DNA, followed by precise breakage and reunion of the phage and bacterial DNA at specific sites.

## Canada Balsam

When specimens are to be made into permanent slides for the light microscope, these materials have been widely used as a mountant. It is a gum dissolved in xylene that is completely incapable of mixing in water, requiring that all material to be mounted must have been previously dehydrated with alcohol.

## Canaliculus

(1) little channel that occurs during bone sequence emanating outwards from every lacuna. The osteocytes in every lacuna conveys cell processes through the canaliculi to form gap intersections with nearby osteocytes. Numerous lacunae with their encased cells and canaliculi go to make up the Haversian sequence of vertebrate bone.

(2) little channel running between hepatocytes of the vertebrate liver. Caniculi combine to form ducts (lastly one large duct- l the bile pipe) through which the secreted results of the liver cells are released to the stomach.

## Canalization

The alleged presence of pathways being developed, regardless of variety in natural or considerably other genetic variables, guarantees the possible rise of a specific standard phenotype.

## Canalizing Selection

Selection method that eliminates individuals who are prone to changing in reaction to variations in environmental influence.

## Cancer

A group of diseases characterized by invasive neoplastic growth at one or more sites and, in many cases, metastasis of cancerous tissue. Cancerous tissues are frequently referred to as malignant tumors, as opposed to benign tumors, which are typically non-invasive. Cancers of various tissue types are known as carcinomas, sarcomas, lymphomas, and so on.

Except for a few exceptions (such as leukemias and retinoblastomas, which are common in children, and teratomas, which are embryonic tumors), most cancers are primarily diseases of old age.

Some cancers, such as Burkitt's lymphoma in humans and Rous sarcoma in chickens, are undoubtedly caused by viruses. However, a viral etiology for all tumors should not be assumed. Cancers are found in eukaryotic fauna and possibly even plants.

## Capacitation

Process that mammalian spermatozoa must go through before they can fertilize, which is normally affected by secretions in the female reproductive tract. The exact mechanism is unidentified, but it is believed to involve the removal of seminal plasma proteins that coat the sperm surface.

This makes it possible for the acrosome reaction to occur and thus fusion with the egg membranes. Capacitation can also be achieved *in vitro* with the use of various chemically defined hypertonic media.

## Capillarity

The peculiarity in which a fluid is drawn upwards on a strong surface or up a fine empty tube on account of the strength of the attractant powers between the fluid and the hydrophilic strong involved.

## Capping

(1) convergence of a particular protein at a limited locale of the outer layer of a cell. The cycle is energy-subordinate and can be shown, by the utilization of fluorescent antisera, to start with summed up restricting to all pieces of the cell surface, trailed by a progression of the bound molecules to form a cap or collection at one area on a superficial level.

(2) one of the posttranscriptional adjustments to a significant part of the hnRNA and mRNA integrated into eukaryotic cells. Covering comprises the expansion of a methylated guanosine buildup to the 5'- end of the RNA.

## Capsid

Many viruses have a protein coat composed of subunits known as capsomeres. When a virus is enveloped by a portion of the cellular plasma membrane, the majority of the capsid coat is contained within the envelope, but some capsomeres may protrude through it. Capsomeres are frequently arranged in the capsid's organization to form an icosahedron or a long filament.

## Carbohydrates

The most plentiful class of natural molecules on the planet earth, carbohydrates are polyhydric alcohols with two hydrogen molecules eliminated to form aldehydes.

The littlest 'unit' of carbohydrate is a monosaccharide having a solitary terminal aldehyde group or a solitary ketone group. Disaccharides have two, oligosaccharides have several, and polysaccharides have numerous monosaccharide units joined by glycosidic bonds.

## Carbon Cycle

The cyclical movement of carbon atoms in nature, which is primarily caused by the metabolism of living organisms. Green plants use atmospheric carbon dioxide in photosynthesis to form complex organic molecules like carbohydrates, and when the plant is eaten by a herbivore, the carbon-containing molecules are modified and used in the herbivore's tissue metabolism.

Similarly, herbivorous animals may be eaten by carnivorous animals, with the carbon being reprocessed. When animals die, their bodies are broken down by autolysis, bacterial and fungal degradation, and much of the carbon is released into the atmosphere as carbon dioxide.

This gas is also produced by the respiration of animals and plants. Furthermore, the carbon sources trapped in fossil fuels such as coal and oil, which were originally derived from plants, are substantially released on burning.

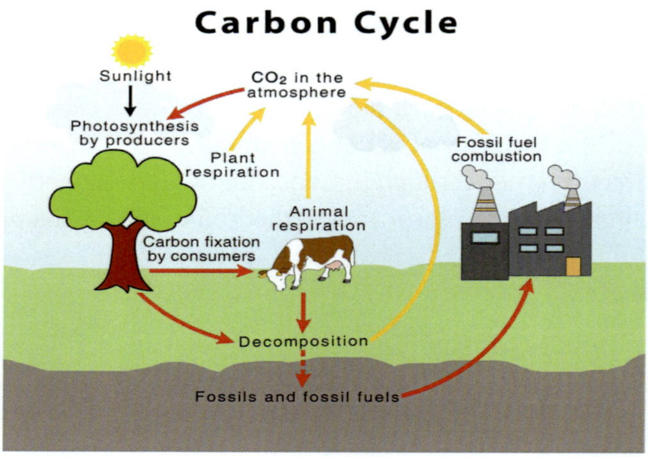


**Figure 7.** The Carbon Cycle.

*Source : https://www.sciencefacts.net/wp-content/uploads/2020/02/Carbon-Cycle-Diagram.jpg*

## Carbon Dioxide

Gas is produced as a byproduct of aerobic metabolism. It is, however, considered necessary for photosynthesis in green plants, so in sunlight, green plants may be net oxygen producers.

## Carbon Dioxide Sensitivity

Some Drosophila genetic strains are unusually sensitive to carbon dioxide and will die when exposed to moderate concentrations of the gas. It has been linked to the presence of sigma virus in the tissues of affected flies.

## Carcinogen

Cancer-causing molecule or physical agent. Many mutagens are also carcinogens, as are benzpyrene, aflatoxins, mustard gas, and agents like x-rays and ultraviolet light.

## Carcinoma

A malignant tumor or neoplasm derived from epithelial or endothelial tissue. With over 90% of human cancers are carcinomas, with 56% occurring in 'external', surface covering epithelia (i.e., skin, gut, and respiratory tract) and 36% occurring in 'internal', glandular epithelia (i.e., Breast, pancreas, thyroid, etc.).

## Carnitine

Carnitine is a vitamin-like compound found in muscle that is responsible for transporting fatty acids across mitochondrial membranes during fatty acid oxidation. Although higher animals can synthesize it, it is a dietary requirement for some insect species.

## Carotenoids

Carotenoids and xanthophylls are photosynthetic pigments found in all photosynthetic plants. Carotenoids are yellow to red in color.

## Carrier

(1) A stable, non-radioactive isotope of an element mixed with a trace of a radioactive isotope to provide a sufficient quantity for chemical handling

(2) Untagged molecules (e.g., non-fluorescent protein) used to bulk up a fluorescent protein preparation.

(3) An individual who is heterozygous for a single recessive gene and therefore transmits the gene to progeny without phenotypic evidence of its existence in his/her genome.

(4) An individual who is chronically infected with a pathogenic strain of a virus or bacterium but does not exhibit disease symptoms. Diseases such as typhoid are frequently transmitted to one another by symptom-free individuals.

## Carrier-Free Radioisotope

Radioisotope sequence to which no transporter has been added; frequently utilized for the most part to portray radioactive material that is of high unambiguous movement.

## Cartilage

Skeletal tissue is made up of chondrocytes, which are groups of mesenchyme cells that secrete an extracellular matrix around themselves. The chondrocytes occupy small spaces in the matrix known as lacunae, and the slightly elastic ground substance they secrete is composed of type II collagen and proteoglycan.

This latter molecule in cartilage is made up of protein bound to the glycosaminoglycan chondroitin 4-sulfate. Cartilage has no blood vessels, so its cells are only kept alive by the diffusion of gases and nutrients through the matrix. The perichondrion is a collagenous connective tissue made up of specialized fibroblast cells in a matrix of type I or type III collagen plus proteoglycan that surrounds cartilage rods.

## Casein

Phosphoprotein is the main protein found in milk. It is precipitated from milk through the action of renin, a process used in the production of cheese.

## Casparian Strip (Casparian Band)

Highly specialized sealing barriers in the apoplast between plant root endodermal cells. These strips work in a similar way to tight junctions between animal epithelial cells through that they prevent solution leakage from the root's central vascular tissue to the outer parenchyma.

## Cassette Mechanism

A genetic mechanism is proposed to explain mating type switching in the budding yeast Saccharomyces. There seem to be two distinct mating types: a and abut cells of one strain could indeed spontaneously switch to the other mating type, a switch that appears to require transposition of DNA sequences within the yeast genome.

## Caste

A group of individuals within a species of social insect (such as ants or bees) which are structurally and functionally specialized for a specific role (e.g., soldiers or workers). The term originated from the strict classes of society into which India was once divided.

## Catabolism

Production of simpler molecules from greater complicated molecules. Catabolism is exergonic and commonly oxidative, giving rise to Ndah whose oxidation is coupled through oxidative phosphorylation to the technology of ATP.

## Catabolite Activating Protein (CAP)

Dimeric protein is composed of two equal subunits (Mr = 2 x 22 500), the product of the CRP gene of *E.coli*. Cap is active solely when the cyclic amp is certain to it when it enables the transcription of positive genes regulated through catabolite repression.

## Catabolite Repression (Glucose Effect)

Effect of glucose in stopping the induction of enzymes required for the catabolism of sugars (e.g., lactose, arabinose and galactose) and positive amino acids (e.g., tryptophan) in *E. coli* and other bacteria. Glucose, or a breakdown product (a catabolite), prevents the expression of the genes for the catabolizing enzymes via regulating the intracellular concentration of cyclic AMP (cAMP).

In glucose-grown microorganisms the level of cAMP is low. cAMP is required to bind to and activate a catabolite activating protein, which in binds to a site on the DNA close to the promoter of a glucose-regulated gene and allows transcription initiation with the aid of RNA polymerase.

Catabolite repression ensures that glucose is used in the choice of inferior carbon substrates (e.g., lactose) which assist in slower growth.

## Catalase

An enzyme is commonly detected in peroxisomes that oxidizes cellular compounds such as phenols and alcohol as well as converting hydrogen peroxide to water and oxygen. The isolation of the powerful and potentially lethal oxidizing agent hydrogen peroxide to peroxisomes, as well as catalase's rapid conversion of this dangerous compound to water and oxygen within these organelles, clearly protects the cytosol from its effects.

## Catalyst

A substance that aids or accelerates a chemical reaction without being broken down in the process. Enzymes function as protein catalysts.

## Cat Box

A consensus sequence of conserved DNA which is most commonly found in approximately 75 base pairs upstream from the initiation site of eukaryotic genes. It is believed to play a role similar to the TATA box. It, like the TATA box, is thought to act as a recognition site for RNA polymerase II enzymes.

## Cathepsins

Proteolytic enzymes are frequently stored within lysosomes. Those very same enzymes play a significant role in cellular breakdown. Cathepsins are in charge of tissue regression or re - modeling during development, metamorphosis, or regeneration.

## Cattanach's Translocation

Translocation in the mouse, B.M. Cattanach discovered a translocation in the mouse involving the insertion of a segment of a chromosome into an X chromosome. It has been discovered that when the appropriate copy of the X chromosome is inactivated, the autosomal genes are also inactivated, albeit partially and depending on the distance from the sex chromatin.

## C Cell

A kind of cell situated in the thyroid and parathyroid organs of vertebrates that secretes calcitonin. In embryonic development, the C cells are derived from the remaining pair of gill pouches.

## Cell

The simple and structural unit of all residing organisms (except a very few acellular examples, e.g., Physarum). The cell structure was once first observed by using Robert Hooke in 1665 for his examination of cork. All cells consist of protoplasm contained inside a surface membrane, the plasma membrane or cell membrane.

They include double-stranded DNA as the genetic material. Although there are many kinds of cells which, via the system of differentiation, turn out to be specialized for precise roles inside an organism. There are three fundamental cell types: prokaryotic or bacterial cell; animal cell and plant cell.

Cells of prokaryotic organisms possess inflexible cell walls outside the plasma membrane and have no distinct nuclei; the DNA is no longer conjugated with histone to structure chromatin. Animal cells are characterized by the presence of a distinct nucleus, the existence of chromatin in which DNA is complexed with histone, the absence of a cell wall at the cell surface and the presence of mitochondria in the cytoplasm.

Plant cells resemble animal cells in possessing nuclei and mitochondria, however, in addition, they have an outer cell wall of cellulose, the presence of plastids in addition to mitochondria in the cytoplasm, and the presence of one or greater massive fluid-filled vacuoles which might also occupy a massive section of the cell space.

Plant and animal cells in multicellular organisms are additionally unique as many or most cells in a plant are in cytoplasmic contact via the plasmodesmata connection, whereas animal cells are related to one any other through specialized contacts - desmosomes, gap junctions and tight junctions.

The common cell is between 1 and 20 f.tm in diameter; the smallest, cells of the prokaryote mycoplasma, are only 0.1 f.tm and the largest single cells are the eggs of birds. Viruses are no longer cellular and should no longer be referred to as living organisms. They are subcellular molecules composed of aggregates of biological molecules obtained from cells.

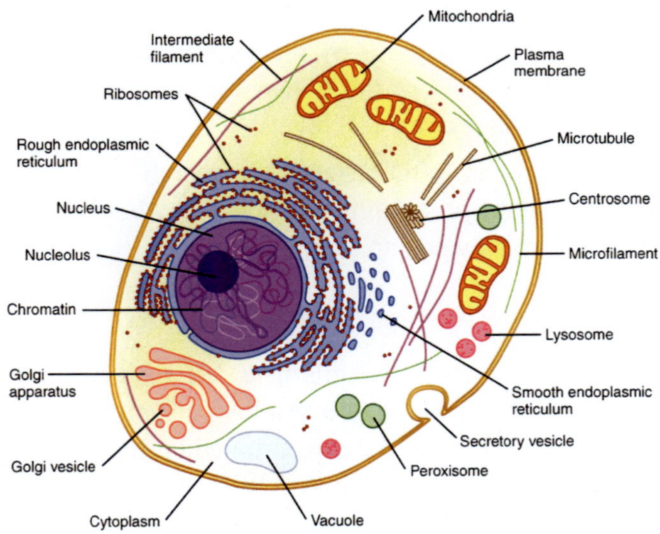


**Figure 7.** Animal Cell And Components.

*Source : Image By Wikimedia Commons*

## Cell Adhesion

The process by which adjacent cells adhere to one another and isolated cells or sheets of cells adhere to specific surfaces. Cell adhesion involves a number of different mechanisms; tissue cells that are somewhat permanently associated are held together by various types of cell junctions.

Fibronectin is an important component of the extracellular matrix that makes it appear to play a role in the adhesion of many tissue cells. Contact inhibition in cell movement in culture is a function of cell adhesion, and the lack of contact inhibition that characterizes many neoplastic or transformed cells is largely due to the absence of adhesion between the cells.

## Cell Aging

As a cell culture ages, the time between successive mitoses lengthens, extending the g 1 phase of the cell cycle. When cell division stops, the cells become disorganized and die. The reasons for cell aging, whether *in vivo* or *in vitro*, are unknown.

Mutations and damage to the cells' DNA may occur, and the effectiveness of repair mechanisms may decline; the efficiency of cellular processes may be inhibited by the accumulation of errors in the machinery responsible for DNA replication, transcription of messenger RNA from DNA, and translation of messenger RNA at the ribosomes; and waste products (e.g., lipofuscin) may accumulate to critical levels.

Furthermore, there is circumstantial research that suggests that differentiated cells may be programmed in some way to divide for a limited number of times before they can no longer divide.

## Cell Aggregation

(1) Formation of clumps of cells in a cell suspension or after cell deposition on a surface as a result of selective adhesion. When cells of various kinds are mixed together, they may showcase a sorting out phenomenon wherein the aggregates are mainly made up of one cell type or another.

2) In the life cycle of some organisms (for instance, Dictyostelium), free-living single amoebae aggregate together, followed by fruiting body formation. Such aggregation is mediated by specific attractant molecules, in the case of Dictyostelium, cyclic amp.

## Cell Compartmentalization

The division of the cell interior into distinct compartments, the majority of which are membrane-bound. They include nuclei, endoplasmic reticula, lysosomes, mitochondria, vacuoles, and chloroplasts in plant cells.

## Cell Coupling

Phenomenon in which adjacent cells may be electrically or chemically connected due to cytoplasmic connections traversing plasma membranes (e.g., plasmodesmata in plant cells and cytoplasmic bridges in some embryonic tissues), but also due to gap junctions something which connect many animal cells and synapses that connect nerve cells.

## Cell Culture Medium (Tissue Culture Medium)

Any of a wide range of media used to sustain animal or plant cells in culture in vitro. An appropriate buffered balanced salt solution, as well as amino acids, nucleotides, and vitamins, is commonly used in certain media. Many cells are particularly choosy and will only grow well in media supplemented with additional components such as fetal calf serum.

## Cell Death

Normal cells die after a limited number of divisions (for example, cultured mammalian lung fibroblasts die after 20-50 divisions), depending on the source animal's age. Cell cultures that have aged take longer and longer to complete their cell cycles until division ceases, the cellular contents become disorganized and clogged with waste materials, and the cells die.

It is unknown why such cell strains die. It has been proposed that as cells differentiate, they may become programmed to die after a certain number of divisions. This number of divisions is known as the Hayflick limit after its discoverer, and the phenomenon is known as programmed cell death.

Cancerous cell lines (for example, HeLa cells) will continue to divide indefinitely., As will some immortalized cell strains that appear to be normal, non-cancerous cell strains. *In vivo*, cell death after a certain number of divisions may ensure that 'rogue' cells that have escaped normal cell division control processes do not proliferate uncontrollably.

## Cell Division

The interaction by which solitary cell divides yields two daughter girl cells. Nuclear division, for the most part including mitosis (albeit in microorganism cell synthesis meiosis is involved), is trailed by division of the cytoplasm and plasma layer to yield the two daughter cells. Following the division of animal cells, a new membrane is synthesized to allow the development that follows, though, in plant cells, cell walls and middle lamella are additionally set down.

## Cell-Free Extract

Fluid extract produced by lysing and homogenizing cells, followed by centrifugation or selective filtration of any remaining whole cells. All of the soluble phases of the cell contents will be retained in such extracts. Depending on the centrifugation or separation conditions, they may or may not contain mitochondria and cell membranes.

## Cell-Free System

Cell-free extracts are used to study aspects of cell metabolism in this synthesizing system. Cell extracts can be used to study the endogenous activity of enzymes or nucleic acids present in the extract, as well as the exogenous activity of protein or nucleic acid molecules added to the system and relying on the activity of processing molecules present in the system.

## Cell Junction

The link between cells. Plant cells frequently have extensive cytoplasmic connections, known as plasmodesmata, and many embryonic cells are linked by cytoplasmic bridges.

Other cell junctions serve as communicating links (e.g., gap junctions, synapses), impermeable junctions that control chemical communication and provide tissue integrity (e.g., tight junctions, septate junctions), and adhering junctions that provide tissue strength and elastic adhesion (e.g., belt desmosomes, spot desmosomes, hemidesmosomes).

## Cell Locomotion

Cell movement caused by cilia or flagella providing motile forces, or by amoeboid movement caused by the complex interplay of cell surface and cytoplasmic microtubules. Differential adhesion between cells is also important in cell movement through and across tissues.

## Cell Lysis (Cytolysis)

Cell rupture is caused by disruption of the enclosing plasma membrane. It can occur as a result of lysosomal activity from within (autolysis), infection of bacterial cells with virulent phage (bacteriolysis), or exposure of cells to other lytic agents such as detergents or osmotic shock.

## Cell-Mediated Immunity

T lymphocytes are the effector agents of the response in one of two adaptive immunity systems (not secreted antibodies). Immunity is generated specifically against intracellular viral infections, cancer cells, fungi, parasites, and foreign tissue. It entails recognizing self-histocompatibility antigens as well as foreign antigens in order to lyse infected cells.

## Cell Memory

Cells retain their commitment to a specific path of specialization (determination) even in the absence of overt differentiation. Memory is thought to be stored in complex feedback loops between the nucleus and the cytoplasm that, once established, are very stable even after many rounds of cell division.

Established lines of cells in tissue culture will retain their differentiated character after many rounds of cell division, whereas insect imaginal disc cells, while embryonic in nature, can be experimentally demonstrated to be highly determined.

# D

## 2-(4-D) (2,4-Dichlorophenoxyacetic Acid)

Synthetic auxin is widely used as a selective herbicide because it is highly toxic to broad-leaved grasses and weeds while causing little harm to narrow-leaved grasses and cereal crops.

## Dalton

The unit of mass used to express molecular masses is the kilogram (usually, but incorrectly, called molecular weights). A Dalton, also known as an atomic mass unit (amu), is defined as 1/ 12 of the mass of one carbon-12 atom and equals 1.663 x w-z g. It is also equal to the mass of one hydrogen isotope atom (hydrogen-1).

Strictly speaking, the Dalton should only be used for molecular mass because molecular weight (m,), also known as 'relative molecular mass,' is a relative (dimensionless) number that requires no units.

## Dark Field Microscopy

A microscope fitted with a special condenser that passes an oblique light beam through the specimen is used in this technique. This light beam is not visible through the eyepiece, resulting in a dark background with light scattered or reflected by interfaces in the specimen appearing as sharply defined bright areas in the field.

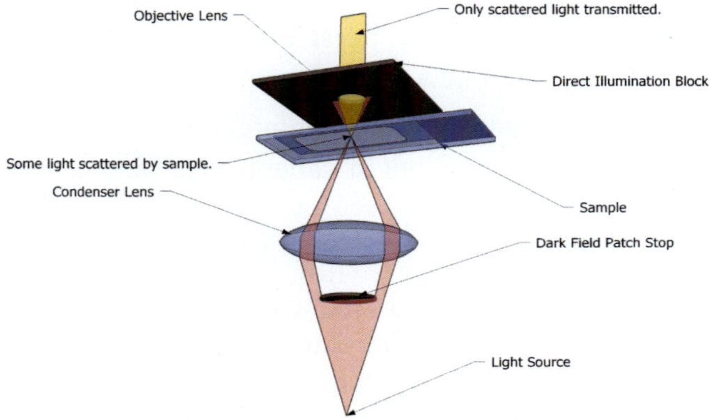


**Figure 8.** Dark Field Microscope.

*Source : Image By Wikimedia Commons*

## Dark Repair (Dark Reactivation)

Repair of DNA upon damage (e.g., by alkylating agents or ultraviolet light) by special enzyme systems that perform optimally in the absence of visible light. This distinguishes these repair enzymes from other polymerases which perform somewhat similar functions but require visible light, a process known as photoreactivation or light repair.

## Darwinian Selection

Selection of organisms within a population based on their superior fitness in comparison to the rest of the population. Thus, natural selection is assumed to achieve 'survival of the fittest.'

## Darwinism

Charles Darwin proposed the theory of evolution with the publication of The Origin of Species in 1859. (His original theory) Individuals with characteristics that make it possible to create more viable offspring than less well-endowed individuals will tend to have their characteristics maintained and spread through a population, according to an impressive body of evidence.

This is known as natural selection; individuals with desirable characteristics were said to be more fit, hence the phrase "survival of the fittest." As the environment changes due to natural events such as climatic variation, new characteristics may be selected, resulting in evolutionary change and the emergence of new species.

Recent evidence suggests that such evolution does not occur very often with time at a constant slow rate, but in rather abrupt saltatory progressions.

## Dauermodification

Phenotypic change in a cell that is induced by that of the environment (and is therefore not genetic), but persists for a time in the daughter cells produced vegetatively. The change will eventually fade away due to dilution. Irradiation-induced damage to mitochondria or chloroplasts may persist as a dauermodification.

## Davson And Danielli Model

A lipid bilayer with the lipid polar groups facing outwards and each polar surface associated with an outermost layer of protein was proposed as a model for the structure of plasma membranes. Although useful at the time, the model was eventually superseded by Singer and Nicholson's fluid mosaic model in 1972.

## Deamination

Deamination of amino acids to ammonia and keto acids is an example of amine group removal. This is an important process when an organism's food supply contains more nitrogen than it requires. The spontaneous deamination of cytosine to uracil that occurs in DNA at an estimated rate of 100 bases per day in every genome is of genetic importance.

There seem to be repair enzymes that replace the uracil produced by this non - intentional deamination. Because uracil does not normally occur in DNA, this repair is possible. Accidental deamination of 5-methylcytosine (a relatively rare base in DNA) causes a significant challenge because the result is normally occurring thymine that is not repaired.

## Decarboxylation

The removal of a carboxyl group from a compound (as carbon dioxide). Typically, enzymes are used to catalyze this process. Histidine decarboxylase, for example, converts histidine to histamine, and glutamate decarboxylase converts glutamic acid to Gaba.

Some decarboxylations are followed by an oxidation (oxidative decarboxylation) (e.g., the conversion of pyruvic acid to acetyl coenzyme a by the pyruvate dehydrogenase complex).

## Dedifferentiation

Cells that have been committed and modified to perform a specific specialized function lose their specialized character and return to a relatively unspecialized structure and function.

Although well-validated examples of dedifferentiation are rare, the production of poorly differentiated cells of the amphibian limb during the formation of a blastema following partial limb amputation is widely reported. Dedifferentiation followed by redifferentiation is more common in plants than in animals and can be seen in callus formation as well as in many cases of plant vegetative propagation.

## Defective Virus

A virus that is unable to complete its replication cycle in the host cell without assistance. This could be due to a genetic mutation that prevents some part of the replicative process, but some viruses are considered defective in the wild-type state because they only replicate successfully when another helper virus is present in the host cell.

Such reliance on a second virus necessitates the use of a biochemical catalyst, such as a polymerase molecule produced solely by the second virus.

## Degenerate Code

Many words (i.e., amino acmes) in DNA and RNA are coded for by more than one codon: for example, both UAC and UAC code for tyrosine. Because there are 64 possible permutations of triplets made from the four different bases, but only 20 amino acids and the 'stop' signal must be coded, degeneracy of the code is permitted.

In most cases, this observation has led to the wobble hypothesis, which proposes that the molecular recognition site on the appropriate transfer RNA molecule, when associating transiently with the appropriate triplet of bases on the messenger RNA, has some inherent imprecision that can compensate for changes in the third base of the triplet.

## Degree Of Freedom

1) A statistical term that refers to the number of distinct pieces of data that can vary independently.

(2) The number of independent variables in heterogeneous (i.e., multiphase) equilibria. Including the number of chemical and physical components (e.g., temperature and pressure).

## Dehydration

Water removal tissues are dehydrated during the preparation of specimens for electron microscopy or permanent light microscopy preparations. The routine method of dehydration is to immerse tissues in increasing concentrations of ethanol until they are immersed in absolute alcohol.

When specimens are immersed in absolute alcohol, they shrink dramatically and lose morphological integrity. The goal of dehydration procedures is to remove water from cells while retaining as much of the cell's morphological structure and substructure as possible.

## Deletion

Loss of a portion of a DNA molecule or chromatin from a larger molecule or chromosome. Individual nucleotide deletions to large chromosomal segments, including many gene loci, are examples of deletions. Loss of a chromosome segment is referred to as intercalary if it occurs within the chromosome or terminal if it occurs at the end of a chromosome.

Most deletion mutations are fatal when homozygous, but many are still viable when heterozygous. In the larva, Drosophila salivary gland cells have both homologous chromosomes lying together, and deletions in the heterozygous condition can be detected by deletion loops, where the nondepleted part of the homolog is visible as an unpaired loop of chromosome opposite to the deleted segment.

## Deletion Mapping

The use of overlapping deletions to determine the position of a gene on a chromosome or genetic map. It is particularly useful for mapping mutation sites within genes. A cross between a point mutant and a deletion mutant tends to produce no wild-type (non-mutant) recombinants when the point falls within the deleted segment; similar manner, no two deletion mutants can start producing recombinants when their deletions overlap.

## Deletion Method

Procedure for isolating messenger RNA (mRNA) specific to a specific gene that involves hybridization of total cellular mRNA to DNA from a specific cell line under favorable hybridization conditions. When the DNA is derived from a cell line with a deletion at a specific locus, the mRNA population that fails to hybridize is likely to be specific for the gene or genes involved in the deletion.

## Delta-Hemoglobin (Hba2)

In healthy adult humans, it accounts for between 1.5 and 3.2 percent of hemoglobin. Hba2 consists of two alpha-globin protein chains and two delta-globin protein chains.

## Deme

A confined population of freely interbreeding organisms. A gamodeme is a population unit; a genodeme is a population with closely similar genetic characteristics; and a cytodeme is a population with closely similar genetic characteristics. Being a group of individuals who have similar cytological properties. A deme is typically much smaller than a species or subspecies.

## Dementia

Higher brain function progresses due to a variety of factors, only some of which are genetic. Some cases of presbyopia appear to be caused by an autosomal dominant gene.

## Denatured Protein

Protein with disrupted secondary, tertiary, and quaternary structures. A denatured protein is made up of polypeptide chains that are randomly dispersed and have not been folded into the precise conformations required for biological activity.

When denatured proteins are returned to physiological conditions, they only very rarely renature spontaneously. Ribonuclease is the best example of protein renaturation after prolonged boiling.

## Dendrite

One of the many short, branching extensions of a neuron, in distinction to the single, lengthy and commonly unbranched axon. Dendrites form synapses with axons of different nerve cells, consequently allowing electrical impulses to be surpassed from cell to cell via the nervous system.

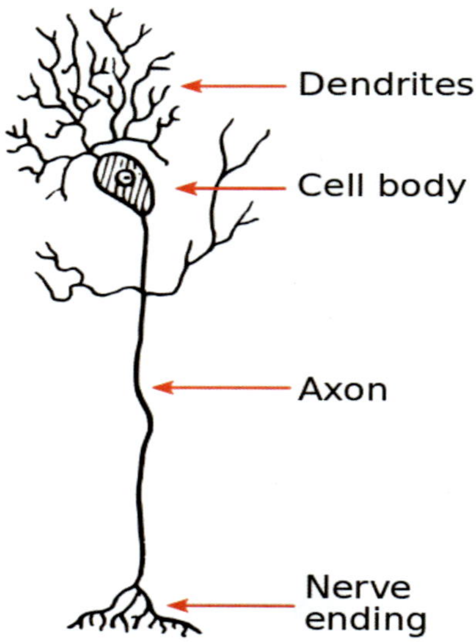


**Figure 9.** Image Showing Dendrites.

Source : Image By Wikimedia Commons

## Dense Bodies

Membrane-bound vesicles in the cytoplasm of animal cells. They are thought to result from the fusion of phagocytotic vacuoles and different pinocytotic vacuoles.

## Density-Dependent Inhibition Of Growth

Reduction in cell division *in vitro* and *in vivo* as a consequence of dense packing of cells due to contact inhibition of motion and finally of cell division. In some situations, poor feedback through particular growth inhibitors referred to as chalones might also occur.

## Dentine

Calcified tissue of vertebrate teeth, lying under the outer enamel and protecting the teeth pulp cavity. Dentine has a comparable composition to bone. Within the dentin matrix is a ramifying network of dentinal tubules containing blood vessels, nerve fibers and the odontoblasts, which synthesize and supply vitamins for the dentine.

In distinction to enamel, which is white and almost completely inorganic in content, dentine is about 75 percent inorganic material and is yellowish in coloration.

## Deoxycholate Detergent

An anionic detergent that is relatively mild and causes minimal unfolding of protein molecules, allowing it to be used to isolate membrane proteins without impairing their activity. Detergents are small amphipathic molecules that can be used to disrupt membranes so that membrane proteins can be isolated.

The detergent molecules' hydrophobic ends bind to the proteins' hydrophobic domains, displacing adjacent lipid molecules and causing membranes to break down into a mixture of water-soluble protein-lipid-detergent complexes and lipid detergent micelles. The detergent molecule's polar end can be non-ionic (e.g., triton) or ionic (e.g., SDS).

## Deoxyribonucleoside

A purine or pyrimidine base is attached to 2-deoxyribose to form this compound. When deoxyribose is phosphorylated, the result is a deoxyribonucleotide.

## Depolarization

Resting potential loss, particularly in synaptic membranes and motor end plates of neurons. Acetylcholine causes this in the neuron by causing sodium and potassium ions to move into and out of the membrane.

## Depurination

Purine residues in a nucleic acid molecule or sample are reduced or removed. It is a common side effect of chemically modifying nucleotide purines (e.g., alkylation).

## Derepression

The removal of a repressor molecule from the sequence or a neighboring control sequence causes the initiation or increase in transcriptional activity of a gene sequence. In the case of induced enzyme synthesis in bacteria, derepression occurs through induction, in which an inducer molecule binds to the repressor molecule, reducing or eliminating its affinity for the operon's controlling operator region.

In contrast to positive control, in which the inducer or other regulatory molecule interacts with the regulatory or operator sequence and positively induces or enhances the activity of the gene or genes, derepression is an example of negative control of gene activity.

# E

## Early Gene

Genes found in viruses' circular genomes such as bacteriophage t4 and sv40. Early genes are transcribed early in the viral replication cycle, and some of their products are required for the transcription or translation of late genes.

Early genes in double-stranded DNA viruses tend to be clustered on one strand and late genes on the other; transcription involves both strands, which also are read in opposite directions.

## Ecdysis

(1) It's the periodic shedding of the exoskeleton by arthropods to allow for growth. Ecdysone regulates it in insects and crustaceans.

(2) In Reptilia, the epidermal layer of skin sheds on a regular basis.

## Ecdysone (Moulting Hormone)

Steroid hormone derived from cholesterol that regulates growth and ecdysis in insects and crustaceans. By puffing on insect salivary gland polytene chromosomes, Ecdysone was instrumental in demonstrating the first hormonal action in stimulating messenger RNA synthesis by specific gene loci.

## Ecodeme

A deme is a local interbreeding population of organisms that is invariably associated with a specific habitat wherever it occurs. An ecodeme, for example, could be limited to sphagnum moss bogs, mango swamps, or the tops of alpine mountains.

## Eco RI Restriction Enzyme

The hexanucleotide target sequence is recognized by type ii restriction endonuclease (DNA restriction endonuclease), and cleavage of the DNA duplex occurs at positions four nucleotides apart in the recognition sequence, yielding fragments of DNA with protruding cohesive 5'-termini four bases in length.

## Ectoderm

A layer of cells that forms the embryo's external covering as well as the neural tube. Nervous tissues, epidermal tissues like hair, horn, and the lens of the eye, as well as all epithelial tissues in sense organs, the anal canal, and the mouth lining, are all derived from embryonic ectoderm.

## Ectopic Expression

Expression of a molecule (for example, a hormone) in an abnormal location. Some tumors, including secondary tumors, produce hormones as an example.

## Ectopic Pairing

Non-specific part of heterochromatin segments in Drosophila salivary gland chromosomes so that adjacent but non-homologous pieces of heterochromatin align and may become physically joined.

## Ectoplasm

The term refers to the surface layers of cytoplasm in ciliates such as amoebas. This outer layer, also known as the plasma gel, is generally homogeneous and devoid of organelles.

## Effective Population Size

The population size of an organism that contributes genes to the next generation. It differs from absolute population size in that individuals lost due to disease, predation, or senility, for example, contribute to the annual total of the latter but may no longer contribute to the former.

## Effector

A nerve cell or organ that reacts to a nervous stimulus. Thus, muscle tissue and glands respond to nervous system stimuli by contracting or secreting.

## Effector Molecules

Small molecules, the majority of which are hypothetical, but some of which have been identified in bacterial gene regulation, play a role in editing the affinity of a repressor or activator molecule for a DNA sequence. Effectors act by combining with the regulatory molecule, thereby changing its ability to bind to the DNA site.

## Egg Apparatus

Structure found inside the undeveloped organism sac of angiosperm ovules. It consists of an egg cell, which produces the embryonic organism following fertilization, as well as a couple of synergies (synergid cells) whose presence is accepted to aid in the fertilization process.

## Egg Membrane

Most animals have a protective and/or nutritive membrane that surrounds the oocyte or fertilized ovum in addition to the plasma membrane. Egg membranes are classified into two types based on their origin.

Membranes of primary eggs. These form within the ovary and are derived from follicle cells that surround the oocyte or from the oocyte itself. The vitelline membranes of amphibians, insects, mollusks, and birds are examples, as are mammals' zona pellucida, sea urchins' jelly coat and fertilization membrane, and the chorion of tunicates, bony fishes, and insects.

Egg secondary membranes- These are derived from the oviduct and other female accessory genital organ secretions that are added to the egg during its expulsion. The jelly coat of amphibian eggs is one example, the mermaid's dogfish purse Bird egg albumen and shell, as well as reptilian egg horny shell.

## Ehlers-Danlos Syndrome

Human disease caused by an autosomal dominant gene (although X-linked and recessive genes are known). It is distinguished by hyperplastic skin, lax joints, and poor wound healing.

## Elaioplast (Oleoplast, Lipidoplast)

It is a type of plant plastid that stores lipids as oil or fat. Because some algal chloroplasts store oil, they are also known as Elaioplast.

## Elastin

Structural protein is abundant in yellow connective tissue fibers in animals. It is highly elastic and is found in tissues that are extensible, such as long alveoli and the walls of large blood vessels.

## Electrochemical Gradient

The gradient determines the electrical potential of the plasma membrane. This gradient may be influenced by both the ion concentration gradient and the membrane potential gradient.

## Electron Acceptor

A molecule that takes up electrons within, the electron transport chain of mitochondrial energy conversion. NAD+ is a major electron acceptor.

## Electron-Dense

In electron microscopy, a region that is opaque to electrons. It could even be interpreted to mean that there are high concentrations of heavy metals present, or that the material in the area has bound heavy metal molecules used for fixation or staining (for example, osmium, lead, or uranium).

## Electron Microscope (EM)

A magnifying microscope that employs electron beams rather than the visible light used in traditional light microscopes. An electron gun emits electrons, which are focused in the evacuated chamber by banks of magnetic lenses.

Because the wavelength of an electron beam is much shorter than that of visible light, the resolving power of an electron microscope is much greater; a good light microscope can resolve less than 200 nm, whereas an electron microscope can resolve less than 0.5 nm. Using a special microtome, specimens for the electron microscope must be cut into ultrathin sections.

MThe material must be fixed and desiccated before staining with an electron-dense stain like uranyl acetate. The standard electron microscope I~ a transmission machine (transmission em): high-voltage transmission em can penetrate thick sections and be used to resolve some three-dimensional structures, particularly when a tilting stage is used.

By reflecting electrons onto a cathode ray tube, scanning EM creates a television-like image. Despite the lower resolution, gross three-dimensional structure and surface features can be seen.

**Figure 10.** Image Showing Electron Microscope.

*Source : Image By Wikimedia Commons*

## Electron-Opaque Tracer

Compound (e.g., metal salt or ferritin) used in the electron microscope to detect the presence of specific molecules or structures in the electron microscope to indicate possible molecular transport pathways (often in association with binding proteins such as antibodies).

## Electron Transport Chain (Respiratory Chain)

Series of electron carriers of regularly increasing redox potential. For example, in the internal mitochondrial membrane and in the cytoplasmic membrane of most aerobic bacteria electrons from NADH flow from a low to an excessive redox potential (02). The electron transport chain that catalyzes this system consists of three protein complexes.

NADH: ubiquinone oxidoreductase, which carries many proteins consisting of these having FMN and iron-sulfur prosthetic groups. The second complex is ubiquinol: cytochrome c oxidoreductase; this includes b- and c-type cytochromes and an iron sulfur center. The 1/3 complicated is cytochrome oxidase, which carries a wide variety of protein subunits such as cytochrome a, cytochrome a3 and two copper atoms.

Hydrogen atoms (electrons plus protons) skip between the first two complexes with the aid of way of the membrane-soluble hydrogen service ubiquinone (and its decreased structure ubiquinol). Electrons flow between the ultimate pair of complexes in the membrane by way of the small electron carrier cytochrome c, which is loosely connected to the outer surface of the membrane.

When electrons flow from NADH to oxygen protons are moved throughout the membrane; this creates a proton motive force that couples the energy accessible from oxidation of NADH to that required for the synthesis of a tp.

In chloroplasts and the cytoplasmic membranes of photosynthetic microorganism, a cyclic electron transport chain includes electrons from chlorophyll, activated by way of light to high potential chlorophyll through the way of an electron transport chain containing quinones, flavins and cytochromes. There is an incredible range of electron transport chains in bacteria.

This is once in a while associated with the range of uncommon substrates being oxidized by means of the chains (e.g., methanol, ammonia, succinate, glucose, nitrite, etc.) or to the electron acceptor which, in anaerobic respiration, may additionally be nitrate (for example) as a substitute of oxygen. Although the specific nature of the elements of these specific electron transport chains may additionally range considerably, they are similar in necessities.

## Electrophoresis

Separation of macromolecules or cells based on their overall charge using an electric field. The separation of macromolecules is done in a supporting gel, usually polyacrylamide gel, and thus the technique is called polyacrylamide gel electrophoresis (PAGE). Gradient gels, in which the concentration of the gel increases while the pore size decreases along the length of the gel, may be used in some cases.

This allows for the separation of substances based on molecular size. The detergent sodium dodecyl sulfate (SDS) could be mixed into the gel. Electrophoresis is used to separate molecules from mixtures (preparative electrophoresis) or to identify molecules.

The latter takes advantage of the fact that molecules migrate at constant rates under defined conditions and can thus be assigned rf values (i.e., the ratio of the distance migrated to that of a dye) to enable identification. To serve as reference markers, proteins or other substances with known molecular weights can be added to the sample prior to separation.

Electrophoresis can be used in conjunction with other techniques. In immunoelectrophoresis, molecules are separated by electrophoresis and then identified by antisera precipitation. Electrophoresis can also be used in conjunction with radiolabeling techniques.

Following separation, the gel can be sliced and scintillation counting used to determine the radioactivity in each slice. Alternatively, the gel slab could be exposed to a photographic emulsion and the position of radiolabeled substances could be visualized using autoradiography.

## ELISA(Enzyme-Linked Immunosorbent Assay)

Sensitive technique designed to detect small amounts of specific proteins via antigen binding. A colorimetric reaction, which is frequently based on a peroxidase-conjugated antibody, used to bind the protein, allows for simple quantification of the results.

## Elution

Removing or collecting liquid from a chromatography column; the eluate is the liquid that might also contain fractionated samples.

## Embedding

The process of preparing biological material for light or electron microscopy. After dehydrating the tissue or organ with alcohol, it is immersed in a material such as hot paraffin wax or a plastic resin (e.g., Acrylamide or epon). After embedding the tissue in molten wax or plastic, the entire mass is allowed to harden or polymerize.

A thorough impregnation of the biological material's cells is required to ensure good preservation and sectioning tolerance. After hardening, the block is sectioned into very thin slices with a microtome and mounted on slides for light microscopy or taken up onto coated metal grids for electron microscopy.

## Embryo

A multicellular organism that develops from a zygote but has not yet hatched, been born, or germinated in the case of plants. Embryos are shielded from the outside world by egg cases, seed coats, or the maternal body, allowing them to develop the rudiments of organs and tissues without fear of disease or serious predation.

Hatching or parturition frequently exposes the young to a harsh environment, necessitating rapid adaptation and use of the complex structures developed during embryonic life. Cell division occurs quickly after fertilization of the egg cell in animals, though if the egg is very yolky, division may be limited to the flat group of cells sitting on top of the yolk.

This early cleavage produces a blastula, which develops further turns into a gastrula Embryology is the study of embryos and the processes that characterized them. Developmental biology, on the other hand, is more comprehensive and often includes studies of fertilization (before the embryo) and the juvenile or larval organism (after the embryo).

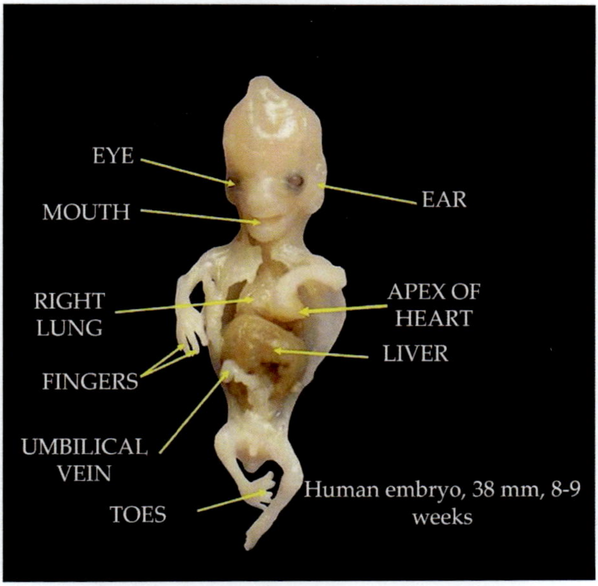


**Figure 11.** Human Embryo (8 Weeks).

*Source : Image By Wikimedia Commons*

## Embryo Culture

Artificial embryo growth through *in vitro* fertilization and subsequent development, or, in more everyday language of the terminology, the aseptic removal of an embryo from its egg case or maternal environment and placement in a special nutrient medium.

## Embryonic Fluid

An embryonic region that exhibits unitary characteristics for a period of development. A field area's distinctiveness has quite often been assumed to result from its common susceptibility to a single chemical or physical factor or force.

## Embryonic Induction

A process that determines the developmental fate of a specific area of an embryo that is at least partially external to the cells themselves. Induced embryonic cells must be capable of responding to the inducer. The nature of embryonic inducers is still unknown.

Some are thought to be simple molecules or specific concentrations of molecules, while others are thought to be physical forces such as light or gravity. In some cases, such as epithelial-mesenchymal interactions, contact with another cell or tissue may be the effective stimulus.

In this case, the underlying mesenchymal tissue instructs the epithelial cells to differentiate in a specific manner. In many of these cases, actual cell contact between the epithelial tissue and the mesenchyme has been shown to be essential, and in the absence of it, the epithelial cells of the mesenchyme simply do not differentiate.

## Emmer Wheat

Plant that was cultivated in the Neolithic period and is thought to be one of the origins of modern wheat. It's a tetraploid wheat variety known as starch wheat or *Triticum dicoccum*.

## Enamel

The exposed surfaces of vertebrate teeth are covered by a whitish layer of calcified material. It is an extracellular matrix composed primarily of inorganic calcium salts in the form of large apatite crystals and is the secretory product of ameloblasts, which are epithelial cells. Unlike dentine, which is produced by neural crest-derived cells, these cells are derived from oral tissues.

## Endocrine Cell

An endocrine gland cell (e.g., the pituitary, adrenal, thyroid, parathyroid and pineal glands, and also the mammalian placenta, the gonads, the islets of Langerhans in the pancreas and parts of the alimentary tract). Endocrine glands secrete hormones into the circulation, where they are carried by the blood to their intended target organs.

## Endocytosis

A method where extracellular materials are brought into cells via membrane-bound vesicles formed by plasma membrane extension or intucking. Endocytosis can be classified as phagocytosis, pinocytosis, or receptor-mediated endocytosis, depending on the nature of the ingested material, the size of the vesicle formed, and the specificity of the process.

## Endoderm

In animal embryos, the innermost layer of cells. During embryonic development, it lines the archenteron of the gastrula and goes on to form the majority of the alimentary tract lining as well as parts of the respiratory tract. The gills and gill pouches, lungs, larynx and trachea, thymus, thyroid glands, tonsils, liver, pancreas, gallbladder and bile duct, urinary bladder, and much of the urinogenital system are all endodermal in origin.

## Endodermis

In plants, the innermost cellular layer of the cortex that forms the boundary between the cortex and the stele. The cells on either side of it differ physiologically, structurally, and functionally. The presence of a band of suberin running around the cell on the radial walls—the casparian strip—is the most distinguishing feature.

The walls of older cells may thicken by becoming heavily impregnated with suberin or lignin. The endodermis is most visible in roots, but it also appears in stems and leaves, especially in pteridophyte stems. A starch sheath forms when there is an accumulation of starch grains in the endodermis.

## Endogenous

Originating within the cell, tissue, or organism as opposed to being introduced from the outside. The terms specific and native are sometimes used interchangeably.

## Endomembrane System

Endoplasmic reticulum, Golgi apparatus, plasma membrane, nuclear envelope, and vacuole and lysosome membranes are taken together. The phrase has occasionally been used to imply a developmental and functional relationship, or even a common evolutionary origin, but this remains speculative.

## Endometrium

The cellular lining of the mammalian uterus, which is often referred to as a membrane despite being more than one cell thick. It is both glandular and mucus secreting, and its structure changes throughout the estrous cycle, which is regulated by estrogens and progesterone.

Its functions during pregnancy are nutritive and protective, and its outer cells are sloughed off as vaginal discharge during menstruation.

## Endomitosis

Chromosome replication without nuclear division, resulting in some degree of polyploidy. Endomitosis occurs in the giant polytene chromosomes found in the salivary glands of *Drosophila* larvae, but the replicate chromosome copies remain attached to one another to form a broad ribbon with homologous chromosome parts in register across the ribbon.

## Endonuclease

Internal DNA molecule breaking enzyme. The enzyme breaks an internal phosphodiester bond, either in one strand only to generate a nick or in both strands to fragment the DNA molecule. Some nucleases (for example, bacterial cell restriction endonucleases), which are now widely used in gene manipulation technology, cut only at specific short sequences of DNA bases.

## Endoplasm

The inner cytoplasm of protozoan cells, as distinguished from the ectoplasm or outer cytoplasm. The cytoplasm of these protozoan cells is frequently layered, resulting in an outer clear plasmagel and an inner opaque plasmasol. The latter also has numerous organelles, including a nucleus, mitochondria, and numerous vesicles and crystalline aggregates.

## Endoplasmic Reticulum (ER)

Reticular membrane system found in the cytoplasm of well almost all eukaryotic cells. It is typically made up of flattened sacs (cisternae) that are interconnected in such a way that the cytoplasm on the inside of the membranous network is distinct from that on the outside.

It is frequently seen to be continuous with the nuclear envelope's outer membrane and, less frequently, with the plasma membrane at the cell surface, and it is also specially modified in certain areas to form a distinct structure - the Golgi apparatus.

In the electron microscope, the endoplasmic reticulum is divided into two types: rough ER, which has numerous ribosomes studded on the outer surface of the cisternal membrane, and smooth ER, which lacks ribosomes.

The rough er is the site of intense protein synthetic activity, with much of the resulting protein being transported or secreted at the cell surface via membrane-bound vesicles that branch off from the main network of the ER. Some cells, such as pancreatic acinar cells, are densely packed with ER cisternae, often in a very regular array, whereas others, such as leaf mesophyll, are noticeably deficient in ER.

The smooth er, which is abundant in muscle cells (and is known as the sarcoplasmic reticulum), is used to store calcium, which is necessary for muscular contraction. Despite the fact that the ER of most cells is strictly confined to the cells, a thin tubular ER cisterna passes through the plasmodesmata to form a continuous bridge between the ER of one cell and the ER of its neighbor.

This is known as a desmotubule. Although it is known that messenger RNA (mRNA) forms polysomes with a number of ribosomes during protein synthesis, the precise architectural sequence of polysomes on the er is unknown, as is the mechanism of the ribosome or mRNA movement relative to the membrane surface.

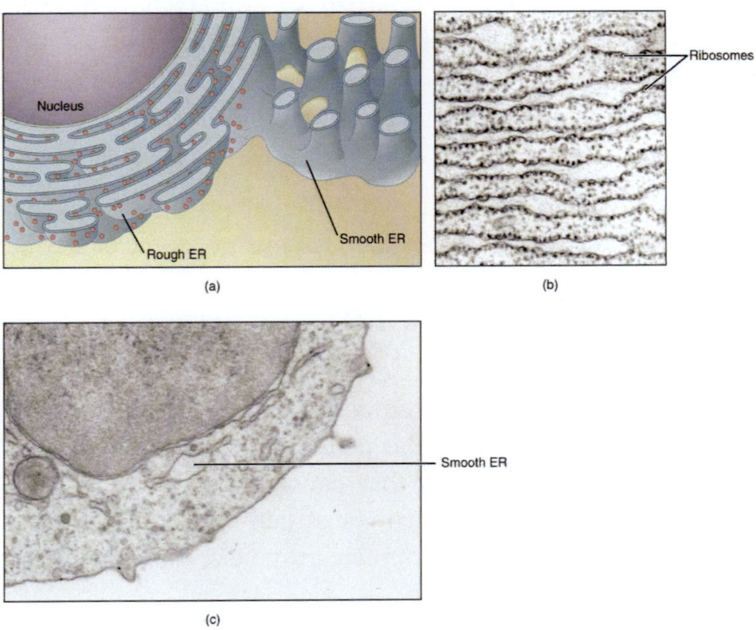


**Figure 12.** Endoplasmic Reticulum.

*Source : Image By Wikimedia Commons*

## Endorphins

A class of large polypeptides isolated from the pituitary gland that are thought to have morphine-like activity in binding to the surfaces of nerve cells in the brain, reducing pain sensation or changing mood. These endogenous opiates are rapidly degraded after release, in contrast to morphine, which appears to owe its narcotic effects to binding to the same receptors for a longer period of time.

There are other endogenous opiates known, such as enkephalins, which have a much lower molecular weight than endorphins but share the same four amino acid sequences that form the presumed binding site.

## Endospore

Spore that forms as a single thick-walled cell within the vegetative cells of certain *Clostridium* and *Bacillus* bacteria. The endospore is extremely resistant to environmental stress and can live for many years. When favorable conditions return, the endospore germinates to produce a vegetative cell.

Endospore production by a bacterial cell is a simple type of differentiation even though special genes are expressed at this time that are not active during the bacterium's life.

## Endosymbiosis

A beneficial relationship between two different organisms in which one organism lives within the cell of the other. As a result, some bacteria exist as endosymbionts in protozoa cells.

## Endotoxin

Toxic substance, usually a macromolecular complex, produced by a bacterium but remaining outside the cell, typically within the cell wall. Endotoxins are found in bacteria that cause dysentery, such as *Vibrio, Shigella,* and *Salmonella,* and injections of cell wall material from these bacteria into mammals cause an immune response, primarily against the endotoxin.

# F

## F1

When two organisms mate, the first filial generation is produced. The parental generation that produced the f 1 generation is referred to as p., and breeding between members of the F1 generation results in F2, or the second filial generation. If both parents are heterozygous for a particular locus, recessive genes are not expressed in the F1 generation, but homozygous recessive individuals can be expected in the F2.

## F2

When two organisms from a F 1 generation mate, a second filial generation is formed. Because some F2 progeny will be homozygous for recessive genes, phenotypic expression of recessive genes can be expected. The phenotypic expression ratios in the f2 can be used to determine the parental generation's true genotype.

## Fab Fragment

A portion of an immunoglobulin molecule that contains the antigen-binding site (fragment antigen binding) and is produced by the enzyme papain. It is made up of one light chain and a portion of one heavy chain joined together by a disulfide bond.

## Facilitated Diffusion

Active transport method in which molecules within a cell's plasma membrane allow small molecules to move through the membrane faster than diffusion would normally allow.

## Facultative

Describes an organism that can live under more than one set of conditions. A facultative anaerobe, for example, can live as an aerobe but will switch to anaerobic respiration if the oxygen level drops. Similarly, a plant is said to be facultatively apomictic if it can reproduce both sexually and asexually depending on the circumstances.

## Facultative Heterochromatin

Highly condensed chromatin that appears to exist only as heterochromatin in specific cells or at specific times. The heterochromatin of female mammals' inactive x chromosome (the Barr body) is said to be facultative.

## FAD (Flavin Adenine Dinucleotide)

Riboflavin derivative. The isoalloxazin ring, which can gain two hydrogen atoms and thus produce the reduced form, is an important part of FAD structure. FAD is the prosthetic group of some membrane-bound flavoprotein dehydrogenases, including succinate dehydrogenase.

It is also the prosthetic group of some oxidases that directly react with oxygen, with the two hydrogen atoms of FADH reacting to form hydrogen peroxide. D-amino acid oxidase and yeast alcohol oxidase are two examples.

## Familial Hyper-Cholesterolaemia (FHC)

It is a genetic disease in which low-density lipoprotein levels in the blood plasma are elevated. It is an autosomal dominant trait that causes atherosclerosis.

## Fanconi's Anaemia

Human disease characterized by extreme chromosome fragility; inherited as an autosomal recessive trait.

## Fascicular Cambium (Vascular Cambium)

Meristematic tissue is found in plant vascular tissue and is responsible for the production of secondary vascular tissue cells when secondary growth occurs.

## Fat (Neutral Fat)

In some plants and all higher animals, this is the primary form of lipid storage. It is deposited in specialized adipose tissue cells and is primarily composed of triglycerides. The majority of storage fat is white fat, but some species have separate deposits of brown fat.

## Fate Map

The fate of the various regions or cells in later embryogenesis is indicated on a map of the egg or early embryo. Pigments are distributed unevenly in the eggs of some species. These can be used as natural markers to track the fates of groups of cells. If no natural markers are present, fate maps can be created using the vital staining technique, which was first introduced by Vogt in 1929.

Small areas of the early embryo's surface are marked with a non-diffusible, non-toxic dye, and the fate of the marked tissues is tracked. These studies enabled a description of the tissues that form in specific regions of the early embryo.

Cells' normal 'fate' can be compared to their embryonic potential; the variety of tissues that can be produced by cells when cultured in isolation or transplanted to a different region of the embryo If a group of embryonic cells has more embryonic potential than fate, then factors must limit cell differentiation during normal embryogenesis.

## Fatty Acid Oxidation (Beta-oxidation)

The process by which acetyl coenzyme an oxidizes fatty acids in animal and plant mitochondria. This metabolic pathway utilizes neutral fat to generate energy for the cell.

## Fatty Acids

Carboxylic acids with a long hydrocarbon chain; lipid constituents. Saturated fatty acids (e.g., stearic acid, CH3(CH2) 16COOH) and unsaturated fatty acids (e.g., oleic acid, CH3(CH2)CH=CH(CH2)COOH) are the two types of fatty acids. Long-chain fatty acids are insoluble in water, but their sodium and potassium salts are widely used in the production of soap.

In the presence of the fatty acid synthetase complex, cells synthesize fatty acids from acetyl CoA and malonyl CoA, with one molecule of acetyl CoA (from mitochondria) linked to two carbons from malonyl CoA.

## Favism

A type of haemolytic anemia (anemia involving erythrocyte breakage) caused by consuming broad beans by individuals who have the genetic glucose-6-phosphate dehydrogenase deficiency.

## Fc Piece

A readily crystallizable portion (fragment crystallizable) of an immunoglobulin molecule produced by papain treatment. It consists of the majority of the constant regions of the two heavy chains connected by a disulfide bond. The Fc piece contains amino acid sequences that determine class specificity, complement binding, and antibody binding to specific cells.

## Fc Receptor

Plasma membrane glycoprotein with a high affinity for the Fc fragment of specific immunoglobulin classes. Fc receptors for iGg are found on macrophages and polymorphs and are responsible for antibody-antigen complex phagocytosis.

Fc receptors for monomeric igg (not complexed with antigen) are found on human syncytiotrophoblast, certain species' yolk sac endoderm, and sucking rodent gut epithelium, where they effect passive immunity transport.

## Fecundity

The number of eggs or offspring that an animal produces. Age-specific fecundity (bx) is defined as the expected number of eggs or offspring produced during a given time interval per female alive at age x.

## Feeder Cell

A cell when added to another cell population in tissue culture, it promotes the latter's growth. Feeder cells are frequently irradiated beforehand to prevent division in culture. The mechanism by which they benefit other cells is due to the phenomenon of metabolic cooperation, which takes place when nucleotides and other molecules pass from cell to cell via gap junctions.

These cells may thus compensate for a genetic deficiency in the recipient cells (for example, a lack of an enzyme required for the synthesis of an essential metabolite), allowing the deficient cells to survive and grow.

## Female Pronucleus

An ovum's haploid nucleus. This nucleus is formed as a result of meiotic divisions that occur during egg formation; the other nuclei formed as a result of these divisions are eliminated or destroyed as polar bodies. The formation of a fertilized egg results from the fusion of the female pronucleus with the sperm nucleus.

## Fermentation

Production of chemical energy in the form of ATP via the degradation of carbohydrates and other organic molecules in an oxygen-free reaction. Fermentation is the only energy-yielding process in obligate anaerobic cells, but it is followed by oxidation of fermentation products in facultative aerobes.

Some yeast and bacterial cells use alcoholic fermentation, which produces ethanol as a byproduct. Glycolysis, which involves lactic acid fermentation, is the most common type of fermentation in animal cells.

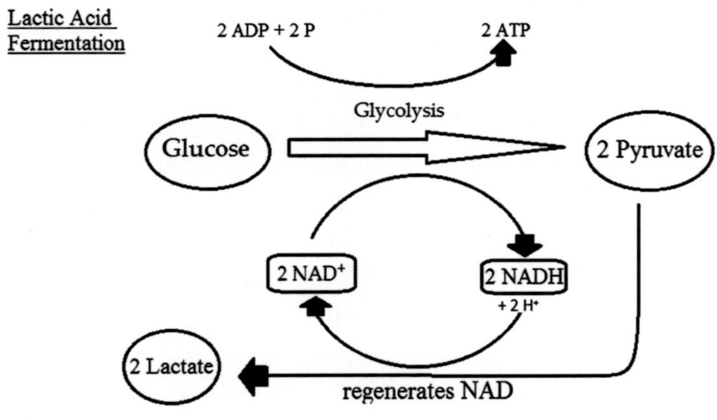


**Figure 13.** Lactic Acid Fermentation.

*Source : Image By Wikimedia Commons*

## Ferredoxins

Best known of the iron-sulfur proteins. They are further characterized as non - haem iron proteins to recognize them from proteins in which the iron is a part of a haem prosthetic group (e.g., cytochromes). They integrate iron-sulfur groups, which have equivalent quantities of iron and inorganic sulfur atoms complexed to the protein via the method of four cysteine buildups.

Treatment with acid prompts the synthesis of a acid labile sulfide. Generally, two (fe2s2) or four (fe4s4) iron and sulfide molecules are available, the iron being tetrahedrally bound to the sulfur atoms. Ferredoxins are normally little, water-solvent proteins that bear single-electron oxidation/reduction responses.

The midpoint redox potentials of the ferredoxins are regularly negative (- 0.4 to - 0.7 v). They work in electron transport chains over respiration, photosynthesis and monooxygenase systems. In anaerobic microscopic organisms, they trademark as the electron acceptor for pyruvate dehydrogenase.

## Fertility

The ability of an organism to produce viable gametes capable of fertilizing gametes of the opposite sex to produce a zygote. Fertility is also entirely compatible with fecundity.

## Fertilization

During sexual reproduction of both plants and animals, male and female gametes fuse to form a zygote. Depending on whether the female gamete, the egg or ovum, is retained within the maternal organism, this event may occur internally or externally. An acrosome reaction ensures that the spermatozoon penetrates the enveloping, non-cellular layer or egg coat in animals (e.g., vitelline membrane, zona pellucida).

Due to the presence of micropyles in an otherwise impenetrable egg coat, access to the egg can only occur at one or several points across the surface of the egg in some systems. A cell recognition system based on species-specific membrane proteins in the plasmalemmae of gametes ensures that fertilization occurs only between gametes of the same species.

In some systems, the point of sperm entry determines the future axes of the zygote, and thus the embryo. Three events occur after the plasma membranes of the gametes fuse and the male pronucleus is released into the cytoplasm of the egg or ovum:

(1) An immediate 'fast block to polyspermy' caused by rapid depolarization of the egg plasma membrane, followed by a transformation of the enveloping layer/egg coat into a fertilization membrane, resulting in a 'late block to polyspermy';

(2) The haploid male and female pronuclei will fuse to restore diploidy and combine the genes from two individuals in a specific way;

(3) Ion fluxes at the cell surface cause changes in intracellular ion concentrations, which activate the egg's developmental program. The unmasking of pre-existing, maternally derived messenger RNA initiates protein synthesis, and DNA synthesis begins.

Non-specific experimental means can initiate egg activation and even very limited early development (e.g., pricking amphibian eggs with a needle). All of these methods are thought to cause ion fluxes comparable to those normally caused by fertilization.

## Feulgen Stain

A crucial cytochemical test for DNA. When positive, it uses Schiff's reagent and produces a purplish-pink color. Microspectrophotometry or colorimetry can be used to determine the DNA content of tissues or solutions by measuring the intensity of the color.

## F Factor (Fertility Factor, Sex Factor)

Plasmid, first discovered in *E. coli*, that transfers itself frequently via conjugation The f factor is made up of a circular DNA duplex (mr5 x $10^7$) that replicates autonomously but in synchrony with the host chromosome.

FIt can integrate into the chromosome, resulting in a variety of HFR strains and promoting chromosome transfer. When separated from chromosome, it may carry one or more adjacent chromosomal genes.

## F' Factor

F factor containing a small segment of the bacterial chromosome caused by abnormal F detachment from the chromosome. Despite the fact that the primary F' cell that produces the plasmid has a defective chromosome, the F' factor is quickly stabilized by transfer to cells with normal chromosomes. To test for genetic complementation, F' factors are used to create merozygotes.

## Fiber

Plant cell that is elongated with tapering ends, a thick secondary wall, and a narrow cell lumen. They are devoid of living cell contents when they reach maturity. Fibers are an important component of sclerenchyma.

## Fiber Tracheid

Fiber found in the xylem with properties intermediate between those of fibers and tracheid. It has bordered pits, like tracheids, but they are usually less noticeable than those of tracheids.

## Fibrin

Insoluble, fibrous protein that aids in the formation of blood clot. The action of thrombin on the plasma protein fibrinogen results in the formation of fibrin. Thrombin is normally produced only in blood that has escaped from circulation and is a product of prothrombin, an inactive precursor found in normal plasma.

## Fibrinolysis

The process of dissolving and dispersing blood clots. In normal circumstances, this degradation involves the action of plasmin on the fibrin of the clot and is rather slow, taking many hours to complete dispersal of even small clots.

## Fibroblast

A connective tissue cell that is in charge of collagen synthesis in cartilage. Tendon, cornea, and so on. Collagen is a polymerized product of tropocollagen, which is synthesized by fibroblasts. Fibroblasts are cells that are flat, elongated, or triangular in shape and are easily cultured in tissue culture.

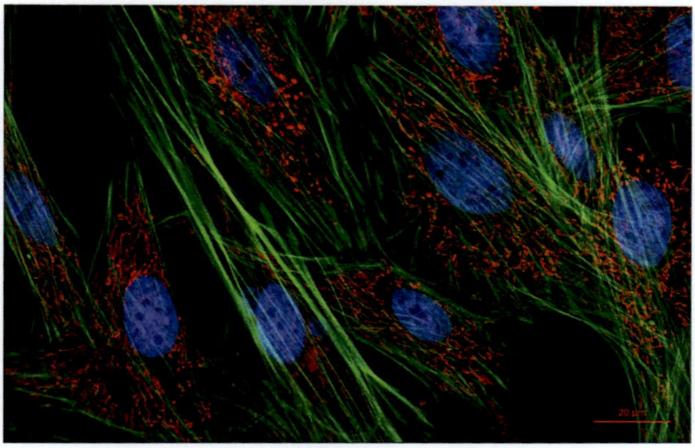

**Figure 14.** Indian Muntjac Fibroblast Cells.

*Source : Image By Flickr*

## Fibroin

Silk is a protein that is the main component of silk, a material used by spiders to make webs and insects to make larval nests and cocoons. Silk is only produced by specialized cells within silk glands and is made up of long chains of the amino acid glycine and alanine with a few serine residues.

## Fibroin Gene

The silk moth Bombyx has a gene that codes for fibroin. It has been extensively studied, as well as remoted and linked in a bacterial clone. It is a massive DNA sequence about 15 kilobases long, with a single intron of 970 base pairs near the gene's 5' end.

Despite the large amount of silk produced by the silk gland cells, there is only one copy of this gene in the bombyx genome. To allow for the massive production of fibroin messenger RNA, the entire genome is replicated 17-18 times in silk gland cells, yielding a DNA content cloth 200 times greater than in a typical diploid cell.

This entire polyploidization process to reap massive output from a single gene sequence contrasts with the gene amplification mechanism discovered in genes coding for ribosomal RNA in Xenopus and egg chorion protein in Drosophila.

## Fibronectin

A class of large proteins secreted by many different types of cells that form a fibrillar matrix that connects cells to other extracellular proteins such as collagen. Certain domains of the fibronectin molecule bind to the cell surface, collagen, proteoglycans, and other molecules.

Cells fail to secrete or bind fibronectin during transformation, which is thought to be responsible for some of the altered properties of transformed cells. When transformed cells are treated with fibronectin, some of their properties are partially normalized. As a result, fibronectin is also known as the lets (large, external, transformation-sensitive) protein.

## Fibrous Protein

One of the fundamental architectural forms used by amino acid chains to build proteins. Fibrous proteins, like collagen, are made up of interlinked chains of polypeptides that form a rope. The main alternative to fibrous protein is packing the amino acid chain into a relatively tight three-dimensional form: such molecules are known as globular proteins.

## Filamin

Long, flexible fibrous protein (Mr- 250 000) that forms dimers in solution. It is important in its role of forming cross-links between neighboring actin molecules, transforming a viscous fluid containing only actin molecules in solution into a solid gel with the addition of filamin: a-actinin is another protein with a similar action to filamin.

## Filopodium

Long, spike-like extension of cell surfaces, larger than a micros pike and oriented in the same longitudinal direction. Filopodia can grow to be up to 50 cm long and are typically found at the growing tip of developing nerve cell axons.

## Filter Hybridization

Exposure of denatured single strands of DNA bound to an immobilizing filter material such as nitrocellulose paper to a solution containing radioactively labeled probe DNA or RNA. Following washing removes only base-paired double-stranded molecules, revealing complementary sequences in the original sample.

## Fine-Structure Genetic Mapping

Analysis of recombination within a gene sequence down to the single nucleotide level. Recombination occurs both within and between gene sequences, and the localization of specific base changes in mutant alleles can be determined by this means.

## Fingerprinting

This technique is widely used to identify differences in amino acid sequence between proteins that are clearly related (e.g., products of mutant alleles).

The technique involves the enzymatic breakdown of the protein into peptides and the separation of such peptides by electrophoresis or chromatography, often two-dimensionally, allowing a comparison of the pattern obtained after migration of the peptides (i.e., The fingerprints).

## First-Order Kinetics

The rate of an enzymic reaction in which the amount of product formed is strictly proportional to the concentration of the substrate rather than the rate of enzyme synthesis or other variable factors.

## Fission

(1) A common method of asexual reproduction in bacteria and protozoa in which a single cell divides to produce two identical daughter cells. Fission typically involves division down the center of the parent cell. Nuclear division takes place first in an amoeba cell, followed by cytoplasm cleavage.

(2) Nuclear fission refers to the division of a eukaryotic cell nucleus.

## Fission Yeast

Yeast replicates by using fission, with the formation of a cell plate in the center of the rod-shaped cell. This is in contrast to propagation by way of bud formation in budding yeast. The best-known fission yeast, widely used in studies on the cell cycle, is Schizosaccharomyces pombe.

## Fitness

The tendency for an organism and its progeny to continue to exist from the possession of a favorable genetic constitution. Fitness is positively selected for in the system of natural selection and is made up of a range of aspects, such as prolonged life expectancy (longevity), the potential to produce numerous offspring (fertility) and the ability to produce exceptionally potential offspring (fecundity).

## Fixation

A system which is used in the protection of biological material for microscopy. It precipitates and denatures most proteins and nucleic acids, thus preventing autolysis. Good fixation, however, aims to cause the minimum disruption to existing biological structures. Fixation also prevents degradation of the material through bacterial decomposition.

## Fixation of Alleles

All members of a population are either homozygous or hemizygous for a specific allele, with all other alleles at that locus having been lost over time.

## Fixation Probability

Probability of an allele becoming fixed in a populace. This likelihood is impacted by various elements, including populace size.

## Fixative

The solution is utilized for the fixation of cells or tissues in the starter steps of their groundwork for microscopy. Normal fixatives are ethanol, acid, formalin and osmium tetroxide. The latter is most usually utilized in the preparation of material for electron microscopy.

## Flagellum

Fine hair-like projection from the surfaces of some cells, with locomotory feature or inflicting motion of the external medium relative to the cell surface. Flagella are observed in some bacteria, many protozoa and algae, and also in sperm. The flagella of bacteria are structurally pretty distinct from those of eukaryotic cells and have a solely locomotory function.

They consist of hollow rods, the walls being composed of a helical array of spherical subunits of a single protein, flagellin. These flagella are primarily based on a rotary motor machine consisting of a protein disc, built-in into the plasma membrane however allowing turning of the flagellar filament.

Eukaryotic flagella are larger than these of bacteria, being about 150 #lm in length and with a structure equal to that of eukaryotic cilia. Eukaryotic flagella, therefore, are in actuality massive cilia, even though because of their giant dimension their movement is wave-like and propagated from base to tip or tip to base.

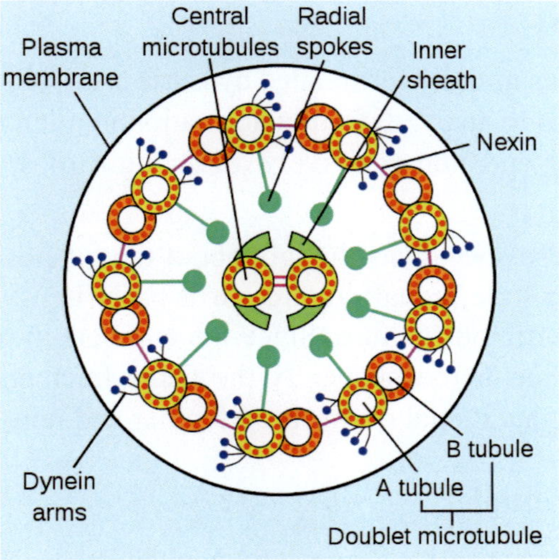


**Figure 15.** A Cross-Sectional Diagram Through A Typical Eukaryotic Flagellum Showing The 9+2 Arrangement Of Microtubules.

*Source : Image By Wikimedia Commons*

## Flame Cell (Solenocyte)

Cell with an excretory characteristic is determined in some low animal phyla (e.g., Annelida and Rotifera). The cells absorb excretory products from surrounding tissues and, via the beating of cilia or flagella, propel these products down a protonephridia tube to the exterior.

## Flanking Sequence

A sequence that is either upstream or downstream, on a similar DNA particle, from a specific authoritative succession like a gene. Enhancer sequences often flank gene coding successions.

## Flavonoid

Plant, blossom, fruit and fungal pigment colors that have a flavone skeleton. Since the biosynthetic precursor of flavone is acetyl coenzyme, flavonoids are included in the acetogenin family of pigments.

## Fluctuation Test

The genetic test was initially formulated by Luria and Delbruck in 1943. The test was formulated to analyze if a transformation in microscopic organisms occured randomly and non-adaptively, or in light of specific ecological demands.

In the test, these specialists thought about the quantities of a mutation occurring in some little, equal cultures of a bacterium with the numbers occuring in one enormous culture. Since the numbers in the small cultures were exceptional factor (consequently the name fluctuation test), the test demonstrated that mutational events are irregular and non-adaptive.

## Fluid Mosaic Model

The model was proposed by Singer and Nicholson, in 1972, to represent the perception of the structure of the plasma layer and negated the previous **Davson–Danielli** model of plsma membrane.

The Singer and Nicholson model suggests that the lipid bilayer of the membrane is basically in a liquid state and that the vast majority of the proteins are situated inside this lipid layer in an intermittent mosaic, either projecting on one surface of the layer or, all the more ordinarily, set between the lipid in order to project on one or the other side.

## Fluorescence-Activated Cell Sorter (FACS)

An instrument that can isolate cells that fluoresce from those that don't as they go down through a restricted opening for examining. Cells might be formed with a particular fluorescein-formed antibody which binds just to a confined populace of a cluster of cells. These will in this manner be chosen by the cell sorter.

## Fluorescence Microscopy

Microscopy permits the fluorescent parts to be envisioned and localized. At the point when stimulated by ultraviolet light, certain synthetic substances fluoresce (for example Emit light of visible wavelength). At the point when sequences of biological material are seen through a magnifying instrument utilizing ultraviolet radiation, fluorescence should be visible where the fluorescing substances are found.

Additionally, fluorescent stains and colors (e.g., acridine orange, fluorescein) can be utilized to name biological components at the subcellular, cell or tissue level.

## FMN(Flavin Mononucleotide)

Subordinate of the b nutrient riboflavin. It is the prosthetic group of the membrane-bound NADH dehydrogenase. A complex flavoprotein is a principal complex associated with the electron transport chain for the oxidation of NADH by oxygen. The hydrogen molecules of FMNH2 are delivered as protons, the electrons being moved to the iron-sulfur center of the enzyme complex.

## Fetal Calf Serum

Extricate from the blood of fetal calves that are generally used to fortify tissue culture media for the development of mammalian cells. The serum contains development factors, some known, others obscure.

## Fetus

Stage in the improvement of warm-blooded animals between when the organs have full-formed and birth. In individuals, the fetal period commences toward the end of the second month of pregnancy.

## Fold Back DNA (Snap Back DNA)

Short sequence sequences of the eukaryotic genome which can ''fold back' on themselves to frame duplexes following denaturation. These sequences can be either scattered or found in clusters.

## Folic Acid

Nutrient of the B complex; an antecedent of tetrahydrofolic acid, a coenzyme with a significant role in purine and pyrimidine synthesis and along these lines in the synthesis of nucleic acids. Its inadequacy is connected to the induction of megaloblastic iron deficiency in warm-blooded organisms.

## Follicle

(1) Vertebrate oocyte with a related layer of follicle cells; relevant all through all phases of oogenesis.

(2) Invagination of the mammalian epidermis into the dermis to produce a pit, containing the hair root and into which the channels of sebaceous glands are unfilled.

(3) Plant organic product shaped from a solitary carpel which, when dry, dehisces along one edge in particular (e.g., larkspur).

## Follicle-Stimulating Hormone (FSH)

Pars distalis secretes Gonadotrophin of the pituitary gland, stimulated by the hypothalamus. This hormone stimulates the growth of graafian follicles in the ovary in the female, however, it stimulates the seminiferous tubules to produce sperm in the male. Follicle-Stimulating Hormone (FSH) is a type of glycoprotein.

# G

## G0

By resting or quiescent cells, phase of the cell cycle entered, which leads to the suspension of division and normal metabolism drastically reduced. From the observation that, the suggestion that cells may be able to enter a resting or go phase stems, in the G1phase of the cell cycle at a crucial point, an enhanced resistance to adverse conditions is displayed by many cells as compared to other stages of the cycle. It is referred to as the r point, when the point in G1 at which such cells can rest and opt out of the normal cycle, then presumed that the cells have entered G0.

## G1

Corresponding to the growth phase, is this phase of the cell cycle.

## G2

When the DNA is replicated, , it is a phase in the cell cycle between the S phase and the M phase, when the cell undergoes mitosis. Thus, In the G2 phase, the nucleus of cells contains twice the amount of DNA (4c, i.e. Four-time haploid) than in the G1.

## Gaba (Gamma-Aminobutyric Acid)

In the brain there is a neurotransmitter, formed by the decarboxylation of glutamic acid, that is used in the inhibitory synapses. The permeability of the postsynaptic membrane to chloride ions is increased by Gaba, so that it becomes hyperpolarized.

## Galactosaemia

This is a human genetic disease inherited is an autosomal recessive disorder. This gene is present on the short arm of chromosome 9. Those who are homozygous for the allele, exhibit enlargement of liver and spleen and some mental retardation. . It causes congenital deficiency of galactosyl-1-phosphate uridyltransferase, and if galactose is removed from the diet, symptoms can be relieved.

## Galactose

Having six carbon atoms (i.e., Hexose), aldose. It is considered as an epimer of glucose. it arises from the hydrolysis of the disaccharide lactose in mammals, which is the major carbohydrate in milk.

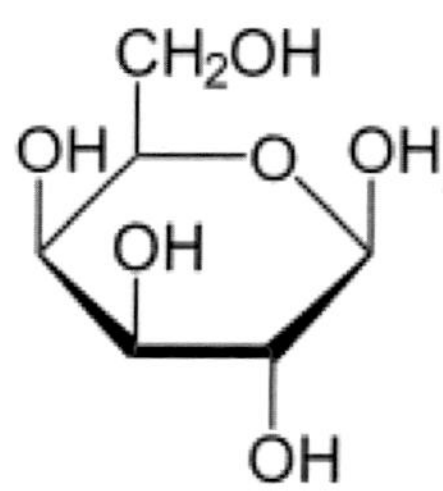


Figure 16. Structure of Galactose

*Source : https://en.wikipedia.org/wiki/Galactose*

## Galactosidase

A second molecule having a hydroxyl group plus an enzyme that hydrolyses galactosides to galactose. It may be methanol (from a methylgalactoside) or nitrophenol (from o-nitrophenylgalactoside) but this is usually another monosaccharide (e.g., glucose, produced from lactose). The b-galactosidase of *E.coli* is an important tool in microbial genetics and the study of the regulation of gene expression. By its natural substrate (lactose) or by a range of galactosides, its synthesis is induced. In extracts or in whole bacteria it is assayed, the hydrolysis of the artificial substrate o-nitrophenylgalactoside to its colored product (o-nitrophenol) is measured. Synthesis of this enzyme laid one of the foundations of molecular biology - understanding of the lac operon of *E. coli* is regulated by the study.

## Galactosyltransferase

From more simple sugars, any one of a group of enzymes that help to make mature complex oligosaccharides. Sugars are finally attached to proteins by many of these complex, and much of this assembly of the golgi apparatus occurred in vesicles. As specific markers for the golgi vesicles these enzymes are used.

## Galton's Apparatus

An equipment in which marbles are dropped through a hole before falling into a series of basal slots and then allowed to bounce randomly between pegs. In the central slot most marbles ended up, but some fall to one side or the other. Piles of marbles in the basal slots reveal a bell-shaped distribution after many marbles have been passed through the hole. For any such series of random events, this pattern represents the distribution probability curve.

## Gamete

WA haploid cell involvedin sexual reproduction. Whence the next generation develops, one gamete undergoes karyogamy/fusion with another to produce a zygote. Ova and spermatozoa are human gametes.

## Gametic Meiosis

For the formation of gametes meiotic division. The organism is diploid for most of its life cycle. This type of meiosis is characteristic of higher organisms, all or a substantial part of the life history is haploid but does not typify organisms.

## Gametocyte

An oocyte or spermatocyte, which, on division, produced gametic cells, either eggs or sperms.

## Gametogenesis

The formation of gametes: sperm and ovum germ cells involved in fertilization.

## Gametophyte

The plant form is composed of haploid cells and produces gametes by mitotic division in part of the life cycle of some plants.

## Gamma-Globin

Along with alpha-globin, a type of globin present in some fetal mammals, including human. In the adult, it is replaced by beta-globin.

## Gamma-Radiation

During the radioactive decay of some atomic nuclei an electromagnetic radiation of short wavelength that is emitted. Radioactive sources such as cobalt-60 are powerful emitters of gamma-radiation and as such may be used to induce cellular mutations or to kill cells.

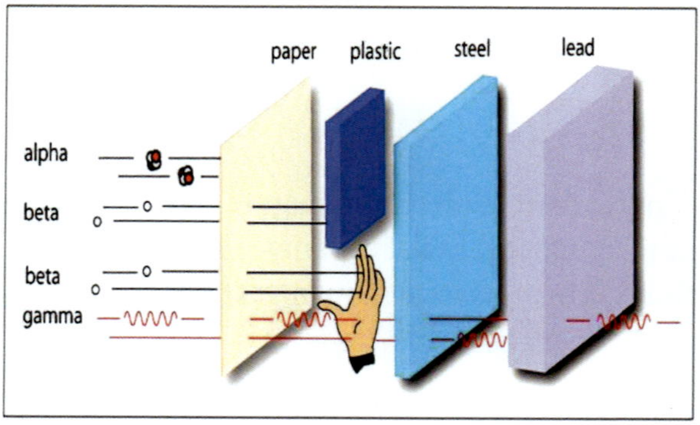


**Figure 17.** An image Showing Gamma radiation.

*Source: Image by ARPANSA*

## Ganglion

An important constituent of the invertebrate nervous system is constituted by a group of neurons. In the autonomic and peripheral nervous system, vertebrates' ganglia occurred, and they also appear in the brain.

## Ganglioside

phingosine, carbohydrate, fatty acid and neuraminic acid contained by one of a group of glycolipids, in nerve cell membranes. In high concentration in the nerve cell membrane of brain cells in patients with Tay-Sachs disease.

## Gap Junction

A common type of intercellular junction creating aqueous channels between adjacent cells, through which small, water-soluble molecules can pass. High intracellular levels of calcium shut down reversibly gap junction channels, it is composed of transmembrane complexes, and connections, each comprising six protein subunits surrounding an aqueous core. There is strong circumstantial evidence for a role in intercellular communication in electrical and metabolic terms.

## Gargoylism

Hunter syndrome and hurler syndrome are two distinct genetic diseases of the human. They result in facial and other deformities; both are connective tissue disorders involving the storage of glycosaminoglycans.

## Gas Glow Counter

A sensitive volume of gas is maintained by allowing constant gaseous flow through the instrument used to measure radioactivity in test samples.

## Gastrin

Cells lining the stomach that secrete Polypeptide hormone. It induces digestive enzyme secretion by other gastric cells.

## Gastrula

The embryo undergoes gastrulation during development. To form the blastula stage, Gastrulation follows the rapid cleavage of the egg into many cells. The formation of the nervous system of the embryo which is initiated in the neurula stage, the gastrula stage precedes.

## Gastrulation

The first and most major morphogenetic movement of tissues in embryogenesis is Gastrulation. Before gastrulation, the embryo has become multicellular, but the cells have not altered their position within the embryo. Cells of the outer layer move into the embryo in an active and coordinated manner during gastrulation. By the invagination of tissue at a particular site in the embryo, the blastopore is formed. Cells are arranged in their germ layers- ectoderm, endoderm and mesoderm- by the end of gastrulation and the embryo has been established which is the basic body plan. Gastrulation allows novel tissue interactions and also brings tissues into contact with each other in new ways. As a result, the alteration of the fate of the dorsal ectoderm to form nervous tissue by its interaction with the newly underlying mesoderm tissue is the most major of these- primary embryonic induction.

## Gated Channel

Depending on reactions with other components, plasma membrane channel of a proteinaceous nature which can be open or closed. Some - the ligand-gated channels - require specific ligand attachment to receptor proteins and other the voltage-gated channels - are sensitive to action potentials.

## Gaucher Disease

A Congenital disease of glycolipid metabolism with high frequency among Ashkenazi jews. It is typified by progressive neurological disease, splenomegaly, bone pain and lipid-filled cells and it is caused by an autosomal recessive gene.

## Gaussian Distribution (Normal Distribution)

As the values deviate more from the commonest or mean value, the pattern of distribution in which the frequency of occurrences of values becomes progressively rarer. When plotted, a symmetrical, bell-shaped curve a gaussian distribution appears. The mean and the standard deviation - a measure of the spread of the data, any Gaussian distribution can be characterized by two parameters. The data to be analyzed are samples from a distribution which approximates to a gaussian distribution, parametric methods of statistical analysis are required. Nonparametric statistical methods must be used if this requirement is not met.

## G Banding

On eukaryotic mitotic chromosomes following giemsa staining, procedure that induces the appearance of distinct bands. Each chromosome is constant and distinct in their banding pattern, unequivocal identification of each chromosome in the set allows. Such banding techniques have unambiguous assignment of gene loci to individual chromosomes and permitted routine assays of human and other karyotypes.

## Gel Electrophoresis

To separate proteins and nucleic acids this analytical and preparative technique is used. The rate of migration is proportional to the net charge and is inversely proportional to the effective size of individual molecules. In slabs or cylinders of a porous, separations are carried out on a semisolid supporting medium such as an agarose gel or a cross-linked polyacrylamide gel. According to the size range of the molecules being separated (for a fixed electric field a higher gel concentration lowers the mobility of a molecule) the concentration of the gel used is varied. Using gradient gels, the best separations are achieved, they have a continuously increasing gel concentration along their length, the best separations are achieved. The molecules can be visualized using a dye on completion of the separation (e.g., coomassie blue for proteins, ethidium bromide for DNA), and to recover individual bands the gel is sliced. If the molecules are radiolabeled, they can be detected by autoradiography. By treating them with sodium dodecyl sulfate (SDS) and 2- mercaptoethanol. Disulfide linkages are cleaved, a useful variation of polyacrylamide gel electrophoresis involves denaturing the proteins, and the through hydrophobic associations, the resulting unfolded protein is coated, with SDS molecules each contributing two negative charges to the complex. The result is that polypeptides of different sizes have the same (or nearly so) density of charge along their lengths (the acquired net negative charge being approximately proportional to molecular weight). On the sieving properties of the polyacrylamide gel electrophoretic separation the smaller polypeptides start moving faster. SDS-page provides a simple method for molecular weight determination as the gels can be calibrated with proteins of known molecular weight.

## Gel Filtration

To separate molecules of different sizes, the chromatographic procedure is used. In the form of small hydrophilic beads in a glass column, the insoluble gel is placed and through the column, a solution containing the molecules to be separated is poured. Smaller molecules enter spaces between or within the beads and are therefore slowed in their progress through the gel but large molecules move quickly. The polysaccharide dextran chains forming the beads provides for the variation in separation characteristics, and different degrees of cross-linking.

## Gel Staining

A staining procedure whereby proteins, nucleic acids or other molecules are visualized in an electrophoretic gel. To stain proteins in gels stains such as coomassie blue are frequently used, by a regulated destaining, until only the desired bands are left stained, the initial staining procedure being followed.

## Gene

Genetic unit consisting of a sequence of DNA, and occupying a specific position or locus within the genome. The phenotype of the organism possessing the gene affected by the gene activity ultimately, but not all genes in the organism that possesses them are expressed. In the genomes of living organisms at least three classes of genes can be identified. Firstly, through the synthesis of an RNA intermediate, the so-called structural genes code for a particular protein molecular sequence and engineer the production of that protein. Secondly, some genes have RNA as a final product but are not expressed as protein. These are chiefly genes coding for transfer RNA and ribosomal RNA. The third type of gene acts as a recognition sequence for other molecules such as DNA polymerases or RNA polymerases or transcription factors and the sequence is regulatory. Most, eukaryotic genes consist of exon coding sequences interrupted by intron non-coding sequences. Some variations in the precise base sequences of an individual gene may exist in a population of such organisms or a diploid organism. Such homologous, but divergent copies are known as alleles. Two differing copies of a gene of a diploid organism are said to be heterozygous for that gene or locus, whereas, for that gene or locus, if the two copies are identical then the organism is homozygous. The genomes of organisms have made it increasingly difficult to provide an entirely satisfactory and unambiguous definition of a gene, there is a great increase in information about the complexity of DNA sequences.

## Gene Amplification

A temporary increase in the number of copies of a particular gene in the genome.. The term is applied to purely temporary increases in gene copy number more frequently, for example in the large ribosomal RNA genes the Drosophila chorion genes in ovarian tissue or in early amphibian development. When such animals are injected with sublethal doses of cadmium, liver cells of mice will amplify the number of genes coding for metallothionein.

## Gene Cloning

In this, individual gene sequences may be isolated from the genome of an organism inserted into viral or bacterial plasmid coning vectors, and thus produced in large numbers.

## Genecology

The population genetics of organisms in relation to their ecological distribution and survival are encompassed by the branch of biology.

## Gene Conversion

It is a phenomenon where alleles are segregated in a 3: 1 rather than a 2:2 ratio, following two rounds of meiotic division. The gene sequences of one allele of a gene have been converted to the sequence of the other allele present in the same cell. The DNA polymerase suddenly switching templates to the other homologous sequence and then switching back may resulted in gene conversion.

## Gene Divergence

The multiple copies of a gene sequence, presumed to have arisen from a single ancestral sequence, form a family of sequences with significant differences in base sequences between the multiple copies is a complete process. For example, human globin genes; the present gene family includes, alpha-, beta-, gamma-, delta- and other globin sequences is presumed to have arisen from the ancestral gene, all having diverged to a greater or lesser extent.

## Gene Dosage

A particular gene is present in a cell or in the genome of an organism with a frequency. For single copy genes such as the fibroin gene in a sperm or egg cell, gene dosage varies from one to the thousands of copies of some ribosomal RNA genes in organisms such as amphibians.

## Gene Expression

In the phenotypic character of an organism, manifestation of a gene or genes. In terms of particular genetic traits, single genes are said to express themselves, as a result of the gene determining the amino acid sequence of a particular protein such as an enzyme, which are most frequent.

## Gene Flow

By interbreeding of individuals in the two populations, movement of alleles from one population of an organism to another. By the new alleles contributing to the range of the gene pool of the population genetic variation within the recipient population is thus increased.

## Gene Frequency (Allelic Frequency)

In a population frequency with which a particular allelic form of a gene occurs. Frequencies of recessive alleles are often much higher than might be assumed from the number of individuals expressing the trait in the homozygous condition. For example, Caucasians in the USA show a prevalence of cystic fibrosis of one in 2000 individuals, but in fact, one in 20 of this population carries the allele.

## Gene Library (Gene Bank)

Collection of cloned genes, frequently comprising most or all of the genes from a particular species. The latter having been made from messenger RNA and lacking intron sequences, such libraries may consist of genomic sequences or DNA sequences. Followed by the cloning of the fragments into viral or bacterial plasmid vectors, libraries are made by cutting purified DNA with selected restriction endonucleases. Using a homologous sequence from a related organism in order to identify the clone within the library which represents the desired gene, gene libraries can be searched.

## Gene Locus (Genetic Locus)

The position of a specific gene on a particular chromosome. Either relative to other genes which are adjacent on the same chromosome, or in terms of a particular mapping distance from the centromere of the particular chromosome locus is expressed as a map position. Thus, a gene might be described as being located on the long arm of chromosome 14 at so many map units from the centromere. Chromosome banding techniques have proved extremely useful in the assignment of genes to particular loci.

## Gene Magnification

The number of copies of the ribosomal genes is increased in mutants of Drosophila since they also have reduced numbers of bristles such mutants are termed bobbed.

## Gene Map (Genetic Map)

From genetic recombination experiments, the order of nucleotides and genes that is deduced. Within a genome or a chromosome as a result of gene mapping, individual genes can be assigned to specific loci.

# H

## Habituated Culture (Anergized Culture)

In simplified requirements for growth that resulted through epigenetic change undergone by plant cells *in vitro* (in suspension or callus culture). As the cells commence synthesis of their own endogenous supplies usually the dependency upon exogenous auxins and/or cytokinins ceases.

## Haem (Protohaem, Ferroprotoporphyrin)

Prosthetic group of the respiratory proteins: hemoglobin and myoglobin and also many cytochromes' respiratory pigments. It is a porphyrin that contains iron (protoporphyrin ix). The free haem group may be oxidized to form haemin which bears a positive charge although it bears no net charge. An iron atom is carried centrally by each haem, each globin chain of the tetrameric hemoglobin molecule carries a separate haem group. It results in the whole hemoglobin tetramer, if oxygen binds to one haem group, an allosteric change is initiated in the haem facilitates the attachment of oxygen to the second haem group.

## Haemagglutinin

Substance that is capable of agglutinating erythrocytes. In their body fluids, invertebrates contain a number of such protein haemagglutinins. Haemagglutinins are usually antibodies of the 1gm class in vertebrates; the human ABO blood group system, among these are the natural isohaemagglutinins.

## Haemerythrin

A respiratory pigment similar to hemoglobin in blood corpuscles of some invertebrates. Hemerythrin has two iron atoms bound directly to the polypeptide chains of the molecule, unlike hemoglobin.

## Haemin (Hemin)

At the center of the porphyrin ring, oxidized form of haem in which trivalent iron replaces the divalent iron atom.

## Haemocyanin

Free in the plasma of some crustaceans, mollusks and arachnids is this copper-containing and blue-colored respiratory pigment found. It will exhibit a bar effect, like hemoglobin.

## Hemocytometer

Cell numbers per unit volume can be counted by a thick glass slide containing a counting chamber of known volume. Including Fuchs Rosenthal, chambers of differing sizes and graduations are obtainable, improved Neubauer, used in general cytological work and frequently used for counting red blood cells.

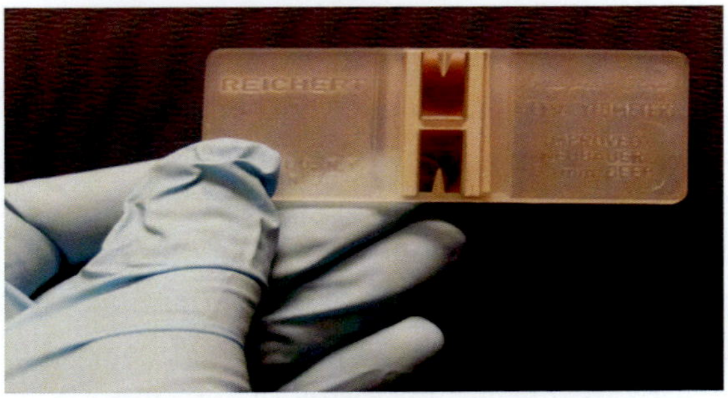


**Figure 18.** A picture of a Hemocytometer .

*Source: Image by Wikimedia Commons*

## Hemoglobin (Hb, Hemoglobin)

A pigmented respiratory protein found in all vertebrate and many invertebrate red blood cells. Consisting of four separate globin chains, each with a prosthetic group of haem attached in most organisms: hemoglobin is a tetrameric molecule. Haem is an iron-containing porphyrin ring, whereas globin is a polypeptide of some 170 amino acids. It has two alpha-globins and two beta-globins in human adults. There being embryonic-, fetal- and adult-type hemoglobin in many mammals, the precise form of hemoglobin present in red blood cells varies with developmental stage. At the center the iron atom of the haem group is able to bind oxygen, to fulfill its function of carrying oxygen from situations of high oxygen tension - the lungs - to situations of reduced oxygen tension- the body tissues, enabled by the hemoglobin molecule. When one or more atoms of oxygen are already bound, hemoglobin acts as an allosteric molecule, having increasing affinity for oxygen. The somewhat bluish color of venous blood is due to the deoxygenated form of the molecule, fully oxygenated hemoglobin is bright red and is characteristic of arterial blood. Genetic aberrations of the hemoglobin molecule are present in diseases such as sickle cell anemia and some types of thalassemia,.

## Hemolytic Disease Of The Newborn

In a rhesus-positive fetus with a rhesus-negative mother occurring condition. Rhesus-negative mothers who have become immunized to rhesus factor (D antigen) make an igg antibody to d which can cross the placenta. By acting as an opsonin, the maternal anti-d destroys the fetal erythrocytes, as D is present on the fetal erythrocytes.

## Hemophilia

A sex-linked disease: results in reduced blood-clotting ability. The single X chromosome present in the male, the disease is only evident in the homozygous condition or the hemizygous situation. Although mating between a hemizygous male hemophiliac and a heterozygous female carrier could yield homozygous females who are hemophiliacs, females are thus normally only carriers. Classical hemophilia, involves a deficiency in the blood-clotting factor viii that is two forms of the disease are known. Type A. About one in 10,000 males are affected when the gene is located at the end of the long arm of the x chromosome. Type B, also called Christmas disease, results a deficiency in the blood-clotting factor ix, is caused by a distinct gene at a locus on the x chromosome distinct from the type a gene, and affects about one in 50 000 males.

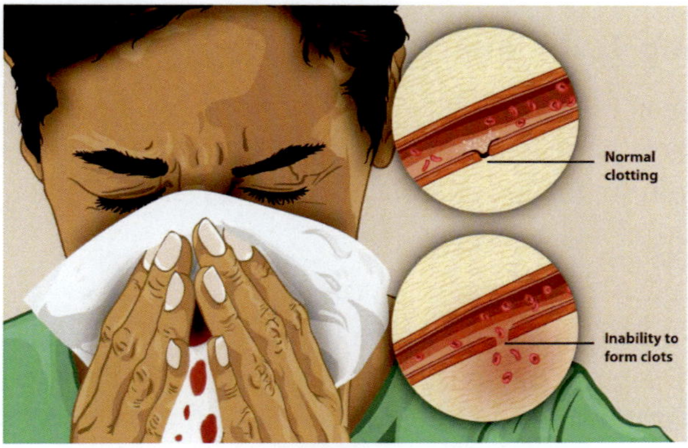


**Figure 19.** A woman suffering from Hemophilia.

*Source: Image by Wikimedia Commons*

## Hemopoiesis (Hematopoiesis)

In the bone marrow and elsewhere by division and differentiation of multipotential stem cells, it is a process by which red and white blood cells are produced. Erythropoietin is a type of hormone, which is produced in the kidney in response to low oxygen tension in circulating blood and stimulates the stem cells to produce more erythrocytes in the marrow.

## Hemostasis

Stopping of blood flow. The term is used: (1) the stagnation of blood within diseased, blocked or constricted blood vessels with reference to; (2) from damaged or severed blood vessels by the contraction of muscles associated with the blood vessels and by the blood clotting processes in connection with the prevention of blood loss.

## Hairpin Loop

By hydrogen bonding between complementary base sequences present in the same strand such double-helical regions of DNA or RNA are formed. Such sequences permit the formation of these loops about the axis of symmetry and exist as inverted repeats of palindromic sequences. With interacting DNA and RNA, these loops may act as three-dimensional recognition regions for molecules. Their existence *in vivo* is hypothetical although such loops are readily demonstrated *in vitro*.

## Hanging Drop Technique

It is a technique in which cells are examined by light microscopy and cultured for short periods of time within a drop of solution suspended by surface tension from the lower surface of a glass slide.

## Haploid

In contrast to the diploid state where two copies of each chromosome are present, describing any organism, cell or nucleus that possesses only one copy of each chromosome. During meiosis, sperm and egg cells are haploid, and become so by reduction division. Some examples of haploids are algae, fungi and protozoa, and the males of some insects such as bees. The n number and the haploid number of DNA as c the haploid number of chromosomes is written. So, for the human n = 23 (46 chromosomes being present in the diploid); a diploid nucleus would be described as 2c. A diploid nucleus after the S phase of the cell cycle is 4c.

## Haplodiploidy

The males are haploid, developing from unfertilized eggs, this genetic system is found in some organisms such as bees, whereas developing from fertilized eggs, the females are diploid.

## Haploidization

It is a process where a diploid nucleus becomes haploid, by a progressive loss of one member of each chromosome pair presumably.

## Haploid Parthenogenesis

There is a situation found, for example, in the honey bee in which haploid cells - the eggs - develop without fertilization into haploid adults.

## Haplont

This occurs in most algae, fungi and many protozoa. Meiosis yields the dominant phase of the life cycle which is haploid and organism in which the diploid phase of the life cycle is the zygote only.

## Haplotype

(1) tend to be inherited together that are set of linked genes (e.g., the a, band c loci of the human hla complex). (2) a single or a few closely linked genes are responsible for a set of isoantigen.

## Hardy-Weinberg Law

Hardy-Weinberg law states that the frequencies of alleles, will achieve equilibrium values in a random mating population of large size in the population. The equation derived from the law, allows the prediction of relative numbers of heterozygotes and homozygotes, based on the Hardy-Weinberg equilibrium. Particularly, if the two alleles are p and q (p+q = 1), the Hardy-Weinberg equilibrium proportions will be p2, q2 and 2pq for the two homozygotes and the heterozygote, respectively.

## Harlequin Chromosome

A cell treated with the thymidine analog 5'-bromodeoxyuridine in a chromosome. In cells that have been grown in the presence of the drug for at least two replication cycles, this compound when included into DNA leads to chromatin that stains less intensely than normal chromatin, chromatids stain differentially and following sister chromatid exchange show a harlequin pattern of alternating dark-and light-stained segments. It can also be used to detect sister chromatid exchange in response to ultraviolet irradiation and other agents the labeling technique and the detection of the resulting harlequin pattern of chromosomes.

## Hayflick Phenomenon

Animal tissue cells will undergo when explanted into tissue culture involved where it is a process in which there is a limit to the number of cell divisions. Cells that deemed to be transformed cells when they grow after the initial 30-40 divisions.

## Heat Shock Proteins

To protect the cell against some of the deleterious effects of heat shock there is a group involved of proteins that are synthesized following heat shocking of a cell or tissue and probably help. Although they are particularly studied in Drosophila probably synthesized in all organisms, where the genes responsible for their production can be seen to show puffing after heat shock.

## Heavy Labelling

In whole cells or in cell-free preparations introduction of a heavy isotope into a compound in order to more easily follow biochemical events involving the compound.

## Heavy Meromyosin

By trypsin one of two active fragments cleaved from myosin. Unlike the light meromyosin chain which is purely filamentous, it contains an ATPase active site and a binding site for actin.

## Heinz Body (Inclusion Body)

Aggregated, precipitated globin molecules, found intracellularly and visible by light microscopy; they are the result of the dissociation of methemoglobin.

## Helix

The solid elements at a constant angle are cut by the (as in the thread of a screw) curve applied to the surface of a cylinder. In the tertiary structure of DNA (which is a double helix) Helices are found in nature, in the alpha-helix which contributes to the secondary structure of many proteins, and in collagen, sperm tails and some filamentous phage many fibrous proteins.

## Helper T Cell

Enhancing the normal activities of these cells, Type of T lymphocytes that functions as a helper cell to other T lymphocytes, B lymphocytes and macrophages.

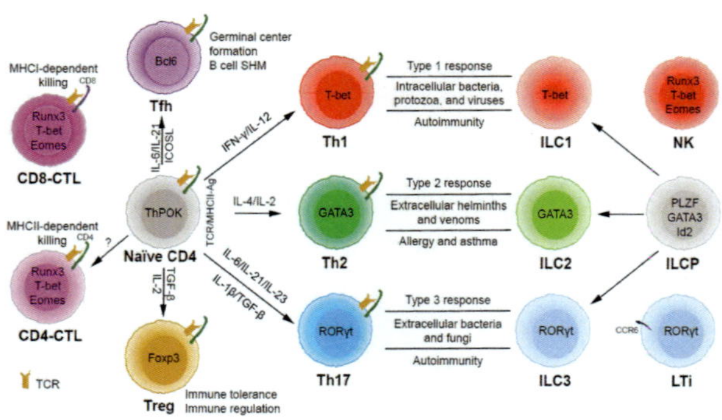


**Figure 20.** An image of Helper T Cell.

*Source: Zhu X, Zhu J. CD4 T Helper Cell Subsets and Related Human Immunological Disorders. International Journal of Molecular Sciences. 2020; 21(21):8011. https://doi.org/10.3390/ijms21218011*

## Helper Virus

Some components required for complete replication and lacks in the defective virus that enables a defective virus to replicate in the host cell.

## Hemicellulose

Act as food reserves in some plants and also a group of alkali-soluble polysaccharides that are important constituents of plant cell walls. They include mannans, glucans, xylans and galactans.

## Hemidesmosome

Occurring on the basal aspect of epithelial cells, cellular structure morphologically resembling half a desmosome and assumed to serve a comparable adhesive function, anchoring the cells to the basal lamina in this case. These may turn out to be minor but recent evidence indicates certain compositional differences from a true desmosome.

## Hemizygous Gene

Usually, a gene on the sex chromosome in the heterogametic sex gene present only as a single dose, (e.g., an X chromosome in an XY male) or as a gene on a piece of a chromosome where the homologous section of chromatin has been lost (i.e., a deficiency heterozygote).

## Heparin

Blood clotting inhibited glycosaminoglycan. The formation of complexes between antithrombin iii allows and activated serine protease coagulation factors, thus irreversibly inactivating those factors. In clinical practice, it is injected as an anticoagulant.

## Hepatocyte

Derived from primitive gut epithelia, the main cell type of the liver. In folded sheets facing the blood-filled sinuses, hepatocytes are arranged. Including the production of bile, they are responsible for the synthesis, degradation and storage of a great range of different substances.

## Hereditary Disease

Usually, a mutant allele of a normal gene, or the lack of a gene or genes, pathological condition resulting from the expression of a gene or genes.

## Heredity

A phenomenon whereby characteristic traits are transmitted from one generation to another. From the transmission of genes from parents to offspring. In the gametes and zygotes, patterns of inheritance within a group of progeny result from the assortment of such genes.

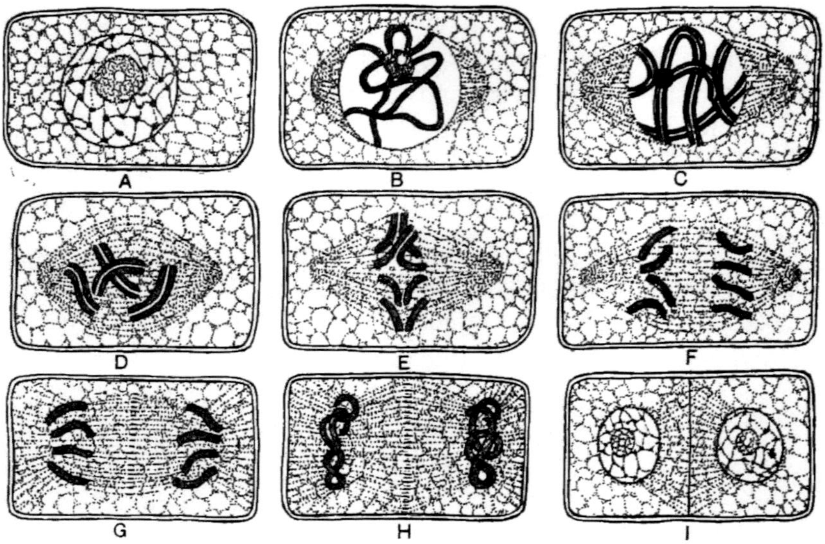


**Figure 21.** An image showing Heredity and evolution in plants.

*Source: Image by Flickr*

## Hermaphrodite

Individuals possessing both male and female reproductive organs. One organ appears before the other in some cases. This is known as sequential hermaphroditism. Other individuals have both ovary and testis at the same time (simultaneous hermaphrodite).

## Herpesviruses

An icosahedral capsid (20 faces) and double-stranded DNA as a genome in animal viruses. In the human herpes simplex virus causes cold sores and genital lesions, and herpes zoster virus causes chickenpox and shingles, herpes-type viruses are cytomegalovirus and epstein-barr virus are examples.

## Heteroallelic

Two homologous alleles are different, describing an individual organism, at differing sites both carrying mutations. Trans test it may be found that intragenic recombination, by applying the cis will result in a wild-type expression.

## Heteroblastic Development

During development changes in the size and shape of an equivalent structure. Young leaves may have a strikingly different shape and form from leaves produced on a more mature plant.

## Heterotroph (Chemoorganotroph)

An organism that obtains its carbon and energy from organic (reduced) carbon compounds. All organisms, except autotrophs, are heterotrophs. They, therefore, include animals, fungi and most bacteria.

# I

## I Band

In light microscope sections of striated muscle fibers, one of the bands with filaments of actin, troponin and tropomyosin which overlap with the myosin filaments.

## Ichthyosis

Extreme piling up of the skin, which is commonly an x-linked recessive trait.

## Idioblast

A cell that is markedly different in function and structure from adjacent cells in the same tissue. (e.g., sclereids).

## Ideogram

By cutting out images of chromosomes from a photograph of a mitotic spread and arranging them in numerical order and homologous pair position, representation of a karyotype in diagrammatic form.

## IgA

It is found in tears, sweat, gut and urinogenital fluids, colostrum and milk immunoglobulin. A dimer of two basic four-chain immunoglobulin units makes secretory IGA. Two alpha heavy chains and two kappa or two lambda light chains are included in each unit. Through their alpha-chains by a cysteine-rich 1 chain, the two units are linked covalently. In the secreting epithelial cells, the IgA dimer is further stabilized by a secretory piece, a single polypeptide chain (Mr=60 000) is synthesized. Aggregated IgA fixes complement by the alternative pathway, IgA provides the first line of defense against invading pathogens, inhibiting their adherence to mucosal cells and so preventing their invasion of the body. In human plasma some monomeric IgA is found. IgA lacks disulfide bonds linking heavy and light chains, two IgA subclasses are known in humans. A secretory immunoglobulin found in birds is also called IgA.

## ID

In several placental mammals, including human the immunoglobulin is found. It has two delta heavy chains and two kappa or two lambda light chains. Its plasma concentration is low (0-0.4 mg/ml). IgD hinge region is extended and this seems to make it prey to proteolysis. Most IgD is found on blood lymphocyte surfaces where, together with LGM, it probably acts as an antigen receptor for the control of lymphocyte suppression and activation. It has a high (13 percent) carbohydrate content. On fetal lymphocytes, high surface IgD levels are found where its presence seems to be associated with antigen priming and increased resistance to the acquisition of tolerance.

## IgE

Found in mammals and possibly birds immunoglobulin (Mr=200 000) at very low concentrations (17-450 ng/ml) in plasma: It is normally bound by its c-terminal end to mast cells (basophils). The n-terminal end of the IGE leads to mast cell degranulation and release of vasoactive amines when contacted with antigen (e.g., histamine). In the gut IgE is putatively a defense against helminth infection, but is principally known for its role in immediate, anaphylactic hypersensitivity.

## IgG

In mammals, it is principal immunoglobulin (m, = 150 000) of the secondary immune reaction; its sedimentation coefficient is 7S; at 3 percent, its carbohydrate content is low. Igg has two gamma heavy chains, each with four domains, three in the constant region and one in the variable region, and two kappa or two lambda light chains. IgG, found at a concentration of 8-16 mg/ml, comprises about 80 percent of the serum immunoglobulin. It is also found in colostrum and milk, IgG is the only immunoglobulin capable of crossing the placenta. They neutralize bacterial toxins and opsonize (coat and make more approachable to phagocytes) particulate antigen, igG antibodies diffuse rapidly into extravascular body spaces. It has four iGg subclasses, designated iGg1_4 and complement by the classical pathway, each with slightly different biological properties. IGg is not found in other vertebrates but is present in all mammals (including the egg-laying monotremes).

## IgM (Macroglobulin)

In mammals, immunoglobulin of the primary immune reaction (m, - 900000), its carbohydrate content is 12 percent and its sedimentation coefficient is 19S. IgMconsists of a pentamer of five basic immunoglobulin units; each unit has two mu heavy chains (four in the constant region and one in the variable region, each with five domains) and two kappa or lambda light chains. By a cysteine-rich j chain, the five units are linked covalently. On combining with antigen, it adopts a crab-like form. But in free solution IgM has a stellate configuration. Although antigen size and occlusion of binding sites mean that this valency is seldom achieved, it has a valency of 10 (i.e., It can combine with 10 antigens). IgG fixes complement by the classical pathway and it is an effective agglutinating antibody. IgM is considered to be evolutionarily the most primitive immunoglobulin and is found in all vertebrates (usually as a pentamer, although monomeric, tetrameric and hexameric IgMs are known).

## IGY

Low-molecular-weight (sedimentation coefficient 7S) immunoglobulin of birds, reptiles and frogs and toads. Its function parallels that of igG in mammals, although igY has a heavy chain rather than a gamma chain.

## Imaginal Discs

As single-cell layers of enfolded tissue beneath the epidermis of holometabolous insects: groups of cells are found (in their metamorphosis, insects have a series of distinct stages). Such as a leg or a wing, each disc represents the undifferentiated primordium of an adult organ, or appendage, and becomes everted at the pupal stage of the adult structure to adopt the complex morphology. Being dedicated to their individual fates without overt signs of differentiation, are described as being determined, by imaginal discs. Trans determination is a phenomenon that involves occasionally changing their fate from that expected of them.

## Immediate Hypersensitivity

To the antigen or allergen, a reaction involving antibodies and occurring within a few minutes of exposure of the individual. To the antigen, individuals displaying such hypersensitivity have normally been previously sensitized.

## Immortalized Cell

By dint of transformation, perhaps by infection with a virus, a cell becomes capable of indefinite growth and division by cells. Immortalized cells are composed of established cell lines in tissue culture.

## Immune Competent Cell (Immunocompotent Cell)

A cell that is capable of responding to an antigen, by division, transformation and production of antibodies or lymphokines of other immunological effector function. In vertebrates, small lymphocytes are normally immunocompetent. B lymphocytes divide and transform into antibody-secreting plasma cells; t lymphocytes divide and transform into effector cells responsible for cell-mediated immunity and t -b cell cooperation when stimulated by antigen.

## Immune Reaction

With antigen measurable or observable reaction or response arising from antibody-antigen interaction, or from contact of immune competent cells. *In vitro* and/or *in vivo*, responses and reactions (e.g., cell division, production of antibody, release of lymphokines and histamine, induction of immunological tolerance and allograft rejection) can occur.

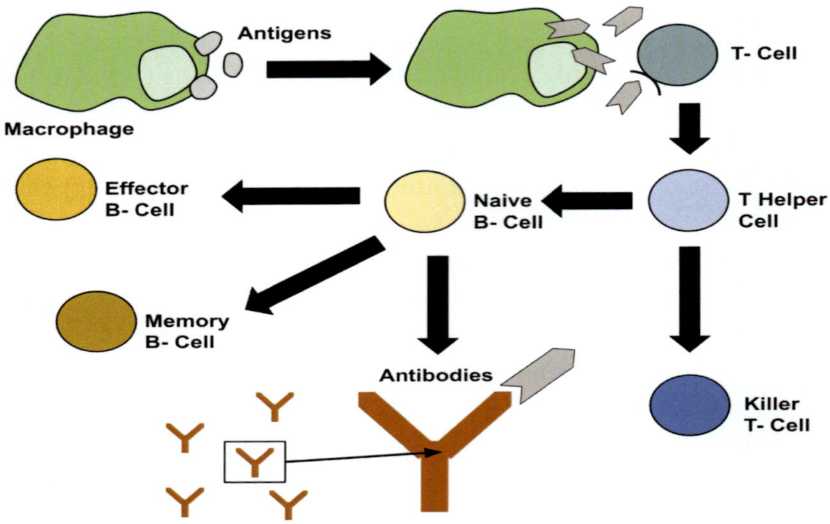


**Figure 22.** A Image showing Immune Reaction.

*Source : Image by Wikimedia commons*

## Immune Surveillance

As a result of continuous monitoring of body cells by the immunological system, a postulated mechanism whereby abnormal cells containing new antigens that have arisen through somatic mutation are removed from the body. It has been suggested that potential cancer cells are removed in this way.

## Immune System

The capacity of an immune response is given by an animal, complex of lymph nodes and, lymphoid cells in the blood and tissues.

## Immunity

The ability of animals and plants to withstand and fight off the invasion by, or pathogenic effects of, foreign organisms and their toxic products is termed as immunity. There are two main types of immunity: non-specific/ innate or acquired/adaptive. By the presence of naturally occurring proteins, in animals, non-specific immunity can be conferred such as interferon and lysozyme and by processes such as encapsulation and phagocytosis. There is a heightened responsiveness to an antigen present on the invasive organism or its toxic products, vertebrate animals additionally display specifically acquired immunity, and which affects its removal more rapidly than would occur in the non-immune state. By humoral immunity specifically acquired immunity is manifested in which circulating antibody is produced, and by cell-mediated immunity in which certain lymphoid cells respond directly to an antigen on their surfaces, through the expression of antibody-like receptor molecules. By antigen and passive immunity, specifically acquired immunity includes active immunity where protection results from antibodies acquired from another individual or species, where an immune reaction is developed following stimulation.

## Immunization

(1) As to confer a state of active immunity or provide passive immunity against infectious organisms or their toxic products, administration of an antigen or an antibody (component of a pathogen) to a vertebrate animal.

(2) For protective purposes, administration of antigen (live or attenuated pathogen) so as to produce an immune response not necessarily (e.g., for structural studies and for other *in vitro* investigations, in order to raise specific antibodies).

## Immunoelectrophoresis

It is a type of technique where proteins are separated by gel electrophoresis and then treated with specific antiserum. This can be used to identify the proteins and determine their electrophoretic mobilities and arcs of precipitated protein form where antigen-antibody conjugation has occurred.

## Immunofluorescence (Fluorescent Antibody Technique)

A method whereby antigens can be localized microscopically in sectioned tissues or in isolated cells in tissue culture and smears. When exposed to ultraviolet light of the appropriate wavelength it involves the use of an antibody labeled with a fluorochrome, usually fluorescein isothiocyanate (fitc) or tetramethyl rhodamine isothiocyanate (tritc) which fluoresces apple green or orange-red, respectively. It involves the application of fluorochrome-labeled antibodies specific for the antigen under study in the direct method of immunofluorescence, whereas the indirect method, utilizes a primary unlabeled antibody followed by a secondary labeled antibody directed against an immunoglobulin, which is more sensitive. Two different antigens to be localized simultaneously by using fitc- and tritc-labeled antibodies specific for one of the two antigens enabled by double immunofluorescence.

## Immunogenetics

Study of the inheritance and genetic basis of the immune response, antibody structure and molecular aspects of antibody-antigen interactions.

## Immunoglobulin Gene

The light chain and heavy chain of the immunoglobulin molecules encoded by the gene. By a series of remarkable genetic mechanisms, the great variability of antibody structure is achieved, which involve rearrangement of parts of the DNA sequence during the life of the B lymphocyte. But in a B lymphocyte they are made adjacent, with an additional J or joining sequence between the variable and constant regions coded separately on the genome. In increasing the sequence diversity of the V region genes, both genetic rearrangement and somatic mutation are involved.

## Immunological Tolerance

While at the same time being capable of responding normally in this way to other antigens, failure of an animal to produce humoral immunity or cell-mediated immunity to a particular antigen. By injecting the antigen during fetal or early postnatal life, immunological tolerance can be induced experimentally, or by injecting adults with very low (low zone tolerance) or very high (high zone tolerance) doses of antigen.

## Immunosuppressive Drug

Drugs to suppress the immune response. To prevent immune rejection of the transplanted tissue, such drugs (e.g., cortisone) are given to patients after transplant surgery.

## Implantation

(1) to a recipient animal, introduction of tissue grafts. (2) to the endometrium of the uterus of the mother attachment of the fertilized ovum or mammalian embryo. Synonymous with nidation.

## Inborn Error of Metabolism

Due to a block in normal metabolism, a biochemical disorder that is genetically determined and produces pathological symptoms.

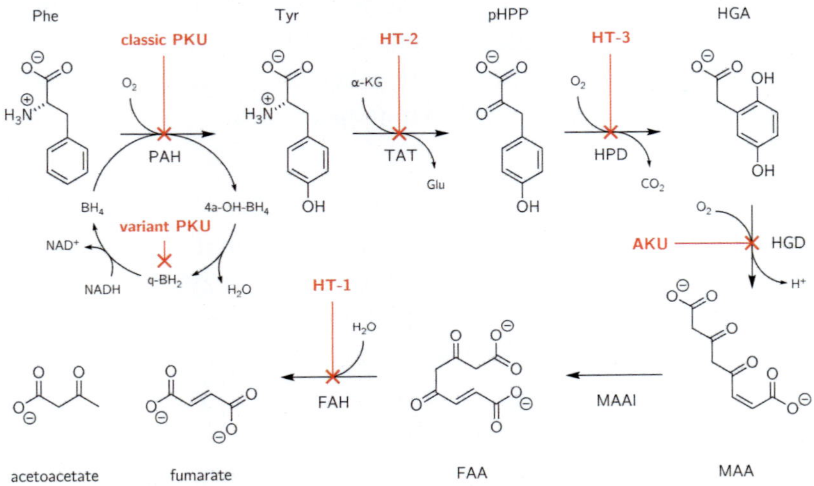


**Figure 23.** Inborn errors of metabolism of phenylalanine and tyrosine.

*Source: Image by Wikimedia Commons*

## Inbreeding

Within a species, breeding of closely related animals or plants, resulting in an increase in general homozygosity.

## Incompatibility

Due to one or both tissues producing antibodies against certain antigens in the other, differences between tissues that result in immunological rejection.

## Incomplete Penetrance

In the phenotype situation in which only a proportion of individuals with a specific genotype express the appropriate characters, usually due to variation in the interaction between the environment and the individual. Penetrance is said to be complete, if all individuals with a dominant mutant gene show the appropriate mutant phenotype.

## Indoleacetic Acid (IAA)

In a wide range of plants principal naturally occurring auxin and one found. Plants that have auxin-like activity have substances other than IAA. Chemically related to IAA, in many cases these substances are indoles. In a number of plant species indoleacetonitrile (IAN) is known to be present, particularly members of the cruciferae, whereas, 4-chloroindoleacetic acid appears to be present as a natural auxin in pea plants.

## Inducer

Usually of low molecular weight, metabolite (or analog of similar chemical structure), which promotes the production of an enzyme, or a series of enzymes involved in other specified proteins or a metabolic pathway. For enzymes inducers are often substrates.

## Induction

(1) by a specific inducer, stimulation or synthesis of a particular product, such as an enzyme. (2) to produce phage following activation of a prophage, stimulation of a bacterial cell. (3) in a tissue determination of a particular developmental or differentiated state by an agent such as another tissue or some physical or chemical agent (i.e., An inducer or evokator).

## Industrial Melanism

In heavily industrial areas prevalence of darkly pigmented forms of animal species. Resulting from the difficulty that predators may have in detecting them against sooty surfaces, it is due to natural selection for mutants that are melanic. In England, the best-known example is afforded by the peppered moth (*Biston betularia*).

## Infection

From the surrounding environment or other organisms, invasion or contamination of an organism by bacteria, viruses or other microorganisms.

## Initial Cell (Initiating Cell)

Plant meristematic tissues stem cells. To make one cell remain meristematic and undifferentiated when such cells divide, they are believed to undergo unequal mitosis, whereas the other will differentiate into a particular specialized cell type.

## Initiation Codon (Start Codon)

A triplet of ribonucleotides in messenger RNA (mRNA) that is at the start of the mRNA coding sequence and initiates polypeptide formation. The triplet is AUG and codes for the amino acid n-formylmethionine.

## Inner Cell Mass

An aggregate inside the trophectoderm of a mammalian blastocyst formed in the group of cells. From the inner cell mass, the whole of the embryo proper is derived.

## Insertional Inactivation

By the insertion of another sequence into it or beside it, is a process, by which a gene sequence is inactivated. To identify phage or plasmid vectors that have accepted copies of novel genes and so have become chimeric, such inactivation is used.

# J

## J Chain

Polypeptide chain (M=35,000) linking, by means of disulfide bonds, the mu chains of the pentameric IgM molecule or the alpha chains of the dimeric secretory IgA molecule. It has a high cysteine content.

## J Segment Of Immunoglobulin Genes

One of three distinct gene segments encoding immunoglobulin light chain (the others are V and C) and of four gene segments encoding heavy chain (the others are v, D and C). The last 10 amino acids of the variable region of light and heavy chains encoded by the J (joining) segments. The random genetic rearrangement between light chain V, J and C, and between heavy chain V, D, J and C create diversity in antibody specificity.

## Jumping Gene

Genetic information coding for and controlling its own transposition carried by the transposon.

## Junctional Complex

On the lateral surface of epithelial cells towards the apical/lumenal end are the assemblage of specialized intercellular adhesions and responsible for: (1) from the lumen sealing the epithelium against leakage of fluids; and (2) the structural integrity of the epithelial sheet maintenance. Typically, it comprises tight junctions (zonula occludens), intermediate junction (zonula adherens) and desmosomes ('macula adherens').

## Junctional Proteins

Association of proteins specifically with cell junctions.

## Junk DNA

Since there may be no positive selection against it, DNA that has no genetic or structural function but persists within the genome.

## Juvenile Hormone (JH)

Determination of whether a molt is to a larval, pupal or adult (imago) form by insect hormone. Molting in insects from one larval stage to another larval stage, or from larval to pupal or adult stages, is induced by two hormones. To secrete a molting hormone ecdysone, neurosecretory cells of the brain produce a hormone that activates the prothoracic gland. By the level of juvenile hormone secreted by the corpora allata, the type of molt is determined. To produce a larval stage; high levels of juvenile hormone cause the molt low levels of juvenile hormone allow metamorphosis to proceed and the molt to produce an adult stage or pupal stage.

# K

## Kanamycin

Antibiotic that binds to bacterial ribosomes and cause misreading of the genetic message involved in the production of proteins

## Kappa Chain (K Chain)

One of two types of light chain found in immunoglobulins which differ by about 60 percent in their constant region amino acid sequences. Both light chains are either kappa- or lambda-type, in any one immunoglobulin molecule.

## Kappa Particles (K Particles)

Within the cytoplasm of killer strains of paramecium, particles are found, and known to represent a symbiotic bacterium *Caenobacter taenospiralis*. Paramecin kills sensitive paramecium placed in the same culture, within the killer strains, a soluble product produced by the symbiotic bacteria.

## Karyogamy

Fusion of two nuclei, normally of gametic nuclei as in sexual reproduction. Fusion of dissimilar nuclei in fungal heterokaryons also falls within the meaning of the word.

## Karyosome

During chromosome pairing, structure in the nuclei of Drosophila oocytes that provides the synaptonemal complexes.

## Karyotype

In a somatic cell nucleus chromosomal complement of a species or individual . To provide a visual record of an individual karyotype, idiograms of numbered metaphase chromosomes are frequently used.

## Kb Cells

In 1955, a cultured cell line derived from neoplastic epidermal cells from the nasopharynx of humans; they were first propagated in a fluid medium by Eagle.

## K Cell (Killer Cell)

Weakly glass-adherent cell with an Fc receptor that potentiates antibody-dependent cellular cytotoxicity (ADCC). In immune surveillance against tumors, K cells may play a role and are arguably the same cells as NK cells.

## Keratin

The structural fibrous protein containing sulfur of horn, feathers, claws, beaks, hair, wool and nails, as well as of the outer layers of the vertebrate epidermis. They are said to become keratinized or cornified when cells accumulate keratin. In the vertebrate genome, a sequence of distinct keratin types is employed in the production of the different keratin-rich components and a large family of distinct keratin genes occurs.

## Keratin Gene

Since keratins are, close to immunoglobulins, the most different proteins, the keratin gene family is extremely huge. The genes range in size from 300 to 3000 bases. What's more, a portion of the more modest ones have no intron. There is an gradient for some inside grouping reiteration inside some keratin gene, and a portion of these inner repeated sequences comprise of inverted repeats.

## Keratin Sulfate

Kind of glycosaminoglycan can be found in the ligament of cornea and the intervertebral plates.

## Ketone Bodies

IBodies formed by the build-up of acetyl CoA which result from unsaturated fat oxidation. They include such substances as b-hyroxybutyrate and $(CH_3)_2CO$. They are formed in the liver and delivered into the blood. Overproduction, known as ketosis, occurs in diabetes and starvation and prompts acidosis and possible coma.

## Ketose

Monosaccharide that contains a ketone group. The least complex model is dihydroxyacetone (a triose having three carbon molecules) and one of the most plentiful in nature is fructose (a hexulose having six carbon atoms).

The c-2 ketone in hexuloses can respond quickly and reversibly with the c-5 hydroxyl group to form an intramolecular ring structure comprising of four carbon atoms and an oxygen molecule (furanose); it is the common type of hexuloses in nature.

## Killer Yeast

Yeast cells containing a double-stranded independent RNA plasmid that encodes genes for synthesis of a stellar substance which is deadly to different cells. The substance is intently comparative in its activity and union to a bacterial colicin.

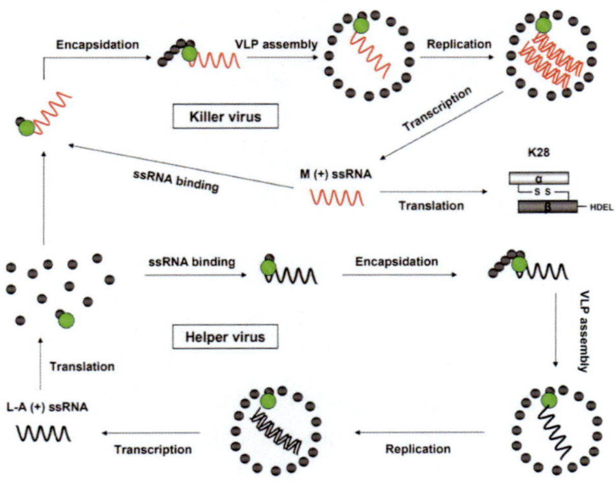


**Figure 24.** Replication Cycle Of L-A Helper And M28 Killer Viruses In *S. cerevisiae.*

*Source : Image By Wikimedia Commons*

## Kilobase (KB)

Unit of DNA length of 1000 nucleotides.

## Kinesis

Development by a cell or organism because of a boost, to such an extent that the rate is subject to the power of the upgrade. Course adjustments of development coming about because of directional boosts are called tropisms.

## Kinetic Complexity

Complexity of a sample of nucleic acids by their reassociation energy in a renaturation experiment.

## Kinetoplast

Particular and exceptionally enormous mitochondrion related to the basal body of the flagellum in some trypanosomes.

## Kinetosome

Basal group of different cell organelles, particularly of eukaryotic cilia and flagella. It is homologous with the centriole.

## Kinety

Line of ciliary basal bodies. The kineties are made out of connected kinetosomes and might be visualized by a silver staining strategy. Kinety design is freely acquired, and when artificially changed by a medical procedure is given in altered structure to progeny.

## Kinins

Group of blood-borne polypeptides that cause smooth muscle withdrawal and vein expansion, the most popular is bradykinin. (2) class of purine derivatives that go about as plant growth regulators, and include zeatin and cytokinin. Manufactured kinins include kinetin.

## Kin Selection

Hypothesis, initially proposed by W.D. Hamilton, that tries to make sense of the advancement of altrusitic way of behaving as coming about because of determination of benefit inside a group of related individuals however not really of benefit to the individual concerned.

## Kirsten Sarcoma Virus

Retrovirus, initially got from mice, however accordingly passage through rodents. It contains the oncogene grouping (v-ki-ras). The human proto-oncogene grouping has not been certainly distinguished.

## Kleinschmidt Spreading Procedure

Procedure created by A.K. Kleinschmidt in which nucleic acids are spread with low quantities of protein, as very thin membranes on cleaned water surfaces. Such films might be gotten on copper matrices (precoated) and consequently analyzed by electron microscopy.

## Klenow Section

Part of the bacterial DNA polymerase I that lacks the exonuclease movement of the enzyme.

## Klinefelter Condition

Acquired human sickness because of a XXY karyotype of the sex chromosomes. Impacted individuals are phenotypically male, yet have little testicles and experience the ill effects of mental hindrance.

## Kornberg Compound (DNA Polymerase I, Pol I)

DNA polymerase found in 1956 by A. Kornberg and colleagues in cells of *E. coli*. It catalyzes the expansion of mononucleotides to the 3'-end of a developing DNA chain *in vitro* with a DNA template, the four deoxyribonucleoside triphosphates and $Mg^{2+}$.

Kornberg compound additionally has a related 3'- to 5'- exonuclease action, eliminating 5'- nucleotides from the 3'- end of a double-stranded DNA molecule (for example: reverse of polymerization).

The pace of polymerization regularly surpasses that of depolymerization aside from when a wrong base is inadvertently included into the developing chain. The two exercises give editing (error repair) aspect in a DNA mix. The chemical has a subsequent exonuclease activity catalyzing 5'- to 3' - denaturation of DNA.

Subtilisin (a protease) separates the Kornberg protein into two unequal pieces. The more modest subunit (mr=34000) holds the 5'- to 3'- exonuclease action, and the bigger (mr= 76 000), called the Klenow subunit, has the 3'- to 5'- exonuclease and polymerization functions.

The movement of Kornberg enzyme to denature DNA in one or the other bearing as well as to catalyze the polymerization response in gap filling and the repair of damaged DNA as opposed to it being the major DNA synthesizing chemical (a job presently credited in *E. coli* to DNA polymerase iii. Pol iii). Without a doubt, bacterial mutants lacking pol I movement stay feasible however have an expanded change rate.

## Kuru

Regenerative infection of the central nervous system in a limited area of New Guinea. It is believed to be brought about by a protein, gained by eating the brains of infected sheep. It is here and there like the scrapie virus of sheep.

## Kwashiorkor

Dietary deficiency coming about because of a lack of a few amino acids, particularly lysine. It occurs in human groups feeding on grain proteins that are lacking in lysine.

# L

## Lactate Dehydrogenase (LDH)

Enzymes that catalyze the interconversion of lactate and pyruvate in anaerobic glycolysis. Lactate dehydrogenase is a tetrameric atom, every subunit of which can be of either the H (heart) or M (muscle) type.

There are five isoenzymes of lactate dehydrogenase (for example Four M-type subunits, three M-type subunits in addition to one H-type subunit, two M-type subunit in addition to two H-type subunits, and so on.).

## Lactic Acid (2-Hydroxypropionic Acid)

Product of glycolysis in anaerobic skeletal muscle during exercise. It is delivered into the circulatory system, thus via the heart, where some might be oxidized, to the liver where it turns into a substrate for gluconeogenesis.

The glucose is delivered into the circulatory system and thereupon to muscle finishes the Cori cycle. Lactic acid is likewise the significant result of maturation by lactic acid bacteria.

## Lactic Acid Bacteria

Microorganisms that produce lactic acid from lactose in milk. They includeee the genera Lactobacillus, Streptococcus and Leuconostoc. They are significant in the development of cheddar and yogurt from milk. The low pH coming about because of lactic acid synthesis forestalls the development of harmful bacteria.

Actic acid microscopic organisms are generally strictly fermentative, without cytochromes.

## Actose

Fundamental sugar in milk; not present somewhere else in calculable sums. It is a disaccharide comprising glucose joined to galactose by a galactosidic bond. The first step in digestion is via hydrolysis to the monosaccharides.

## Lacuna

Space inside a tissue or cell: for instance, the spaces in bone system in which the bone cells occur, or, in plants, water-or gas-filled holes that might give lightness.

## Laemmli Loops

Loops of DNA seen when chromosomes are seen by electron microscopy following removal of histone; named after the originator of the procedure, U.K. Laemmli. The circles all stretch out from a focal lattice or system of nonhistone protein, and each circle is a particular DNA domain.

## Lagging Strand

Strand of DNA that is made discontinuously during DNA replication, and comprises a progression of Okazaki fragments ligated together.

## Lag Phase

The initial period during cell culture development, following inoculation of cells into a new development medium. During this time, practically zero expansion in cell number occurs, yet the cells are to a great extent involved in adjusting to the new medium. The lag stage is ordinarily followed by an exponential development stage.

## Lambda Chain (A. Chain)

One of two kinds of light chains found in immunoglobulins which contrast by around 60% in their amino acid grouping.

## Lambda Phage

Two-fold individual DNA phage that infects *E. coli* and is generally utilized as a cloning vector. Lambda phage might enter either a lytic development or lysogenic development cycle contingent upon an equilibrium of genetic and molecular elements inside the cell.

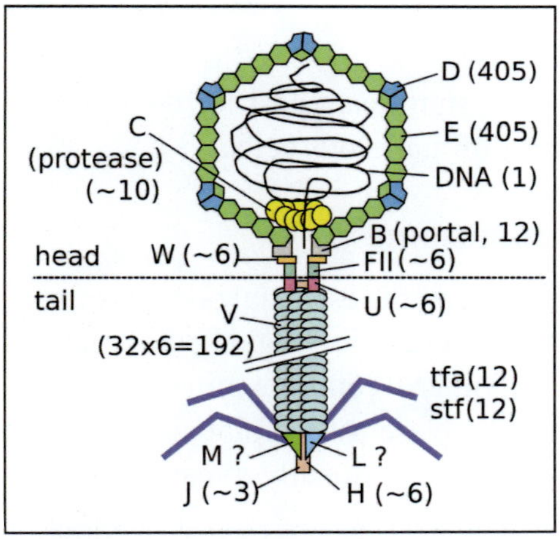


**Figure 25.** Bacteriophage Lambda Virion (Schematic).

*Source : Image By Wikimedia Commons*

## Lamella

(1) Layer/sheet of calcified extracellular network adding to the appositional development of bone; lamellae are stored either sporadically in woven/ cancellous bone, or in concentric layers around Haversian channels in smaller bone. (2) Elective term for one of the spore bearing 'gills' of the sporophore of a basidiomycete fungus (e.g., an agaric). (3) Gelatin rich layer between the paired cell walls of two adjacent plant cells.

## Lamellipodium

Cell surface expansion as a smoothed sheet which is normal for cells like fibroblasts, particularly when motile on a surface under *in vitro* culture conditions.

## Lamina

Levelled part of a leaf upheld by the petiole. In animal cells the basal lamina is an extracellular network discharged by numerous cells and forming sheets between cell layers or tissues.

## Laminin

Glycoprotein found as a significant part for the basal lamina, which forms the extracellular matrix of numerous cells.

## Lamp-Brush Chromosome

Sort of chromosome found in the essential oocytes of vertebrates and seen as enormous quantities of matched circles along the length of the chromosomes, giving them, in the light microscope, a similarity to victorian lamp-brushes. These chromosomes, which are truly diplotene stage chromosomes, remain for a long time and are transcriptionally dynamic.

This component, which is exceptionally uncommon in chromosomes and is shared simply by the polytene chromosomes of Drosophila and other flies, has prompted the lamp-brush chromosome to be an object of extraordinary cytological and genetic interest. The chromosome circles are the decondensed parts of the chromatin and are the sites of transcriptional movement.

## Lanthanum

Component utilized, normally as its hydroxide, as an electron-thick tracer in electron microscopy. The little size of the atom permits it to be utilized to demonstrate destinations of layer penetrability.

## Larva

Immature type of numerous organisms. Albeit generally equipped for taking care of hatchlings are not reproductively dynamic.

## Late Genes

Genes that are expressed late in the life cycle of a virus. These genes as a rule code for proteins like capsid subunits.

## Latent Images

Picture created on the visual film following exposure.

## Latent Virus

Virus that can remain for significant stretches in the host organism without delivering any obvious outcomes of its presence . Such viruses are herpes simplex and chickenpox virus, the last option every now and again becoming inert and repeating later as shingles (herpes zoster). This might be broken by UV or by the presence of a partner virus.

## Late-Imitating X Chromosome (Barr Body)

Dense duplicate of the x chromosome of well evolved organisms. As contrasted and the non-condensed duplicate of the x, this dense duplicate, which is made out of facultative heterochromatin, is replicated late in the S period of the cell cycle.

## LAV (Lymphadenopathy Virus)

Causative agent of human diseases, and presently more commonly known as HIV (Human Immunodeficiency Virus).

## L Cells

Refined cell line, initially portrayed by early in 1943. He took typical subcutaneous connective tissue cells from sound C3h mice and treated them with 20-methylcholanthrene. The cells accordingly created could be refined and yielded sarcomas when infused into solid C3h mice.

## LD50

Lethal dosage 50%; the portion of a toxin (e.g., a toxic substance, drug or ionizing radiation) that will cause 50% mortality in a populace of organisms or cells presented to the toxin.

## Lead Citrate

Heavy metal salt utilized as a stain in the sequence of examples for transmission electron microscopy.

## Leader Sequence

Area at the 5'- end of a messenger RNA that is regularly untranslated yet contains promoter signs associated with ribosome control.

## Leader Sequence Peptide

Short length of amino acid sequence at the amino end of some protein molecules, which, because of hydrophobic amino acids, helps the protein through the membrane of the endoplasmic reticulum thus guaranteeing its possible destination. The leader sequence peptide is usually eliminated before the protein becomes active.

## Lectins

Carbohydrate binding proteins or glycoproteins that agglutinate cells as well as bind complex sugars. Lectins are normally occuring compounds isolated especially from higher plants and seeds, yet additionally from parasites, microbes, kelp, sponges, mollusks, fish eggs, body liquids of invertebrate organisms and lower vertebrates and from mammalian cell membranes.

Despite the fact that their exact physiological job is unsure, lectins have demonstrated valuable in a wide assortment of uses *in vitro*, especially in blood group and erythrocyte poly-agglutination studies, lymphocytes isolation, fractionation of cells, and the isolation and refinement of starch containing molecules.

## Leghaemoglobin

Hemoglobin-like iron containing protein in root nodules of leguminous plants. These nodules are brought about by disease with the nitrogen fixing bacterium rhizobium. In spite of the fact that rhizobium respiration requires oxygen, the presence of oxygen annihilates the nitrogenase and therefore the nitrogen fixing ability.

Oxygen bound to the leghaemoglobin is made accessible to the microscopic organisms while keeping the oxygen pressure of the nodule tissue extremely low. Leghaemoglobin is coded for by a plant gene.

## Leishman's Stain

Stain, utilized in a solution of methanol, to give a blue tone to cell nuclei. It is utilized most often to stain the nuclei of leucocytes.

## Leptotene (Leptonema)

The first phase of meiotic prophase. During this stage the chromosomes show up as thin strings with condensedchromomeres.

## Lethal Mutation

Mutation that causes the death of the organism. Assuming the transformation is latent, organisms that are homozygous will be killed, yet in the event that prevailing, heterozygous individuals will likewise be killed. A few deadly mutants die as embryos and are seen as deviations of Mendel's ratios.

## Leucine (Leu)

One of the 20 normal amino acids. It has a non-polar sidechain.

## Leucocytes (White Platelets, White Blood Corpuscles)

Nonexclusive name for a heterogeneous group of blood cells, known as the white blood cells. These type of cells includes everything other than erythrocytes and blood platelets.

Chiefly leucocytes work as a feature of the body's immune system by: (1) non-specific means (for example phagocytosis of unfamiliar matter); (2) specific means (for example Synthesis of specific antibodies as a feature of the immune system). Leucocytes include granulocytes (basophil, eosinophil and neutrophil), lymphocytes and monocytes (macrophage precursors).

## Leucoplast

Dry pro-plastid, normal in non-photosynthetic plant tissues and from which different types of plastid might be formed.

## Leukemia

A disease described by an overabundance of white platelets (leucocytes). Numerous leukemias are known to be virus actuated, and are somewhat normal illnesses in a wide scope of warm-blooded organisms.

It is additionally a not rare aftereffect of exposure to radiation. One type of the virus in the human - chronic myeloid leukemia - is portrayed by the presence of an additional a peculiar chromosome, the Philadelphia chromosome.

## Leydig Cell

Cell of the mammalian testis that secretes steroid chemicals and is especially rich in smooth endoplasmic reticulum.

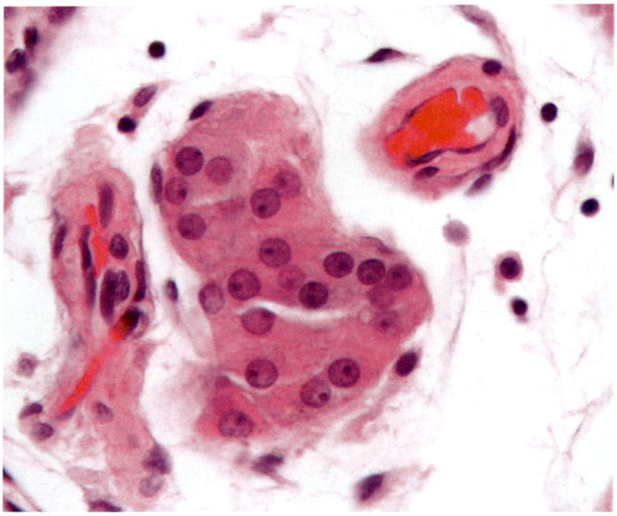

**Figure 26.** Leydig Cells At High Magnification.

*Source : Image By Wikimedia Commons*

## L-Structures

Slow-developing variations of a few types of microscopic organisms (named after the lister establishment). They come up short on typical cell wall, and accordingly look like protoplasts and spheroplasts, however have fortified cell layers. L-structures can return to typical vegetative cells.

## Ligand

Atom whose three-layered structure empowers it to tie with high effectiveness to a reciprocal site on another particle. For instance, a compound substrate is a particular ligand for the chemical.

## Light Chain (L Chain)

More limited of the two kinds of polypeptide chain (mr-22,000) present in all immunoglobulin atoms. Two light chains are connected, generally covalently, with two weighty chains to form the essential immunoglobulin atom.

Human light chains have a variable region containing the n-terminal portion of the chain, most likely engaged with antigen-restricting, and a steady area involving the c-terminal portion of the chain.

Two classes of light chain are found in individuals: kappa (60% of light chains) and lambda (40% of light chains). Kappa and lambda chains can be connected to any weighty chain class, however inside a given immunoglobulin atom just kappa or lambda chains will be available.

## Light Green

Color utilized in organic histology to stain specificly cellulose (which is become red by the stain). It is dissolvable in ethanol or clove oil.

## Light Meromyosin

Little sub-molecular types of muscle myosin. Four molecules of two unmistakable sorts of light chain meromyosin partner with the tops of the two heavy meromyosin molecules to form the functional complex.

## Lignin

Normal constituent of plant cell walls; significant part of wood. It is a complicated polymer of fairly factor organization, made up from alcohols, sugars, sweet-smelling amino acids and phenolic substances.

## Line

Group of homozygous individuals that structure an unadulterated reproducing populace particular from others of similar species.

## Lineage

(1) succession of development from specific genealogical animal groups through intermediates to later and unmistakable animal varieties. (2) series of cells that are connected by a continuous grouping of cell divisions, and in

which the cells at one or the opposite side of the series might show particular differentiative attributes.

## Linkage

Genetical association over inheritance of at least two nonallelic genes. Since individual sequence regularly guarantees that non-allelic genes have a low rate of linkage, genes are supposed to be connected when the genes for linkage are higher than these.

The linkage values got for specific genes are made sense of by the way that they are on a similar chromosome (in this specific situation, here and there alluded to as a linkage group).

Genes near one another on a similar chromosome are supposed to be firmly connected; assuming genes are approximately connected, it suggests that getting over may decrease the degree of linkage recorded albeit the two genes are on a similar chromosome.

## Linker DNA

(1) fragment of DNA between nearby nucleosomes. Its length is variable, averaging around 50 base matches, and it. Is a likely connection site for histone Hl. (2) DNA grouping utilized in recombinant DNA innovation to combine two other DNA sections or the end of one part.

## Linking Number

Number of times that two strands of a double-stranded helical molecules cross each other. Molecules of DNA that are indistinguishable in all regards other than the linking number are named topological isomers.

## Lipases

Enzymes that digest/degrade fats to yield acids and glycerol.

## Lipid Annulus

Ring of lipid atoms reinforced by hydrophobic communications around a natural layer of protein. This lipid might have a distinct e structure from the rest of the layer lipid.

## Lipid Bilayer

Each of the three classes of lipid found in organic layers (phospholipids, glycolipids, cholesterol) contain amphipathic molecules and, as a grouping, all natural membranes are essentially lipid bilayers with related protein atoms.

In fluid arrangements such lipid, atoms will self-gather into active, steady, bi-layered designs by which the hydrophobic tails of the two layers of lipid particle project inwards, sandwiched between the hydrophilic ends which face outwards on every part of the bilayer.

In such a bilayer, sidelong dispersion of individual atoms is very normal and fast (giving ease inside the plane of the bilayer) however flip-flop movement is incredibly intriguing.

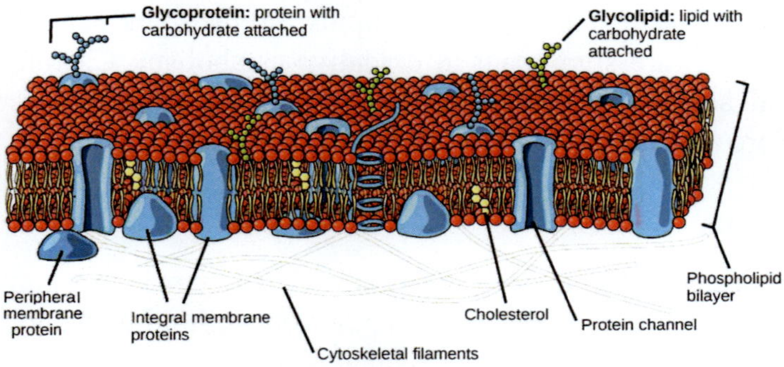


**Figure 27.** Lipid Bilayer With Various Components.

*Source : Image By Wikimedia Commons*

## Lipid Micelle

Amphipathic lipid molecules will not just self-collect into lipid bilayers in fluid arrangements, yet additionally into micelles which are little (colloidal-size), round, sub-molecular totals in which the hydrophobic tails project inwards and the surface contains the hydrophilic groups.

## Lipids

Enormous group of natural mixtures found in living life forms, and recognized by their dissolvability in non-polar solvents like ether and ethanol. They include fats, waxes, phospholipids, glycolipids and steroids, as well as straight-chain hydrocarbons.

## Lipofuscin (Age Stain)

Stain that will in general amass as reddish colored granules in the cytoplasm of enduring, non-isolating cells; granules are in many cases limited by a solitary membrane. Lipofuscin, got from lipoproteins, might be a build up from lysosome movement, especially under hypertoxic conditions. High levels of lipofuscin may debilitate cell functioning and add to cell maturing.

## Lipoic Acid

Compound comprising of unsaturated fat chain spanned by a disulfide linkage and fundamental for starch digestion in certain microorganisms.

## Lipolysis

Synthesis of unsaturated fats from lipids in the presence of lipases. Unsaturated fats are significant in oxidative catabolism. Chemicals (e.g., ACTH and adrenaline) increment lipolysis from fat tissue accordingly making more unsaturated fat accessible in the blood plasma.

## Lipoproteins

Group of formed proteins occuring in blood and lymph that comprise of lipid formed to an amino acid chain. They work in lipid transport between destinations of fat deposits liver, and small digestive system.

## Liposome

Artificial membranous vesicle produced using lecithins and different lipids that is promptly consumed by living cells. Liposomes have been taken advantage of for the conveyance of novel genetic material to cells as an option in contrast to mechanical infusion.

## Lipotrophin (Lipolytic Molecule, LPH)

Peptide created by the mammalian pituitary organ that animates lipolysis.

## Lipovitellin

Lipoprotein constituent of amphibian yolk platelets. Alongside phosphorvitin, it is a cleavage result of a huge precursor protein vitellogenin.

## Liquid Scintillation Counter

Electronic instrument used to gauge amounts of radioactive substances by examining the discharge of radiation during radioactive decay. The sample containing the radioactive compound is put in a liquid containing a fluorescent molecule.

This molecule emanates photons of noticeable light when presented to beta-molecules or gamma-radiation. Photomultiplier tubes measure the light produced thus deciding the radiation levels.

## Locus

Site in the chromosome set of an organism in which a specific gene is ordinarily found.

## Lod Score

LOD (short for "logarithm of the odds") score is a statistical estimate of the relative probability that two loci

## Lomasome (Plasmalemmasome)

Extracellular structure found in some plant cells. Furthermore, particularly in contagious cells. Comprising of membranous vesicles arranged between the cell wall and the plasma membrane (or here and there implanted in the actual wall). It is differently proposed that these designs might be storage sites, or destinations of enzymes associated with cell wall change of amalgamation.

## Long Period Interspersion

Example of conveyance of DNA sequences found in the genomes of certain living beings like Drosophila. It includes non-repetitive DNA sequences between extended lengths of moderately repetitive DNA.

## Long Terminal Repeat (LTR)

Grouping of a few hundred bases found at one or the opposite side of some DNA molecules, particularly the genomic DNA of retroviruses. The repeats at either end are indistinguishable from each other or, in any event a few cases, are transformed repeats.

## Loop

Single-individual area toward the end of a nucleic acid (in which the equal piece will be hydrogen bonded as a duplex) and consequently addresses the bonding between two terminal palindromic sequences.

## Low-Density Lipoprotein

Plasma lipoprotein that is the central source of cholesterol for cells outside the liver and digestive system. Cells bind the lipoprotein through specific receptors, and there is proof that individuals experiencing the acquired problem hypercholesterolemia have nonappearance or lack of these functional receptors.

This inadequacy creates unusually elevated degrees of circulating cholesterol and low-density lipoprotein, with subsequent statement of cholesterol in blood vessel plaques and enormously expanded risk of coronary illness.

## LSD (Lysergic Acid Diethylamide)

Stimulating medication that increase the level of serotonin in the brain.

## Luciferin

Compound responsible for the brilliant (light-emanating) properties of numerous life forms.

## Lucke Virus

Herpes virus that causes adenocarcinomas of the kidney in frogs.

## Luria and Delbruck Test

Experiment in genetic genes intended to show whether a particular change could be natural or happened randomly. It included the utilization of a fluctation test, in which various colonies of microscopic organisms were all the while evaluated for protection from lysis by bacteriophage.

On the off chance that change to phage resistance were random, significant vacillation in the quantities of mutant microbes would be normal in the equal colonies, contingent upon definitively when the transformation had occured (assuming it occured in the the way of life, numerous progency would have collected, if late, no progency would have gathered).

Since checked changes in the quantities of resistant colonies were gotten, it was inferred that transformation was occuring at a steady random rate and was not being prompted by exposure to the phage or other ecological elements.

## Luteinizing Hormone (LH, Interstitial Cell-Stimulating Chemical, ICSH)

Glycoprotein chemical (mr=34,000) made out of an alpha and a beta chain and discharged by the anterior pituitary organ.

In female warm-blooded animals, LH bind receptors that create on thecal and granulosa cells to achieve the development of the graafian follicle, resulting ovulation and improvement of the corpus luteum. In males (where it is known as ICSH) it follows up on interstitial cells of the testis (Leydig cells) making them secrete testosterone.

## Luxury Gene (Tissue-Specific Gene)

NA grouping that isn't expressed in all cells. Such genes include numerous which have been seriously investigated, since extravagance genes frequently code for proteis in specific tissues.

Globin genes, fibroin genes, insulin genes and crystallin genes are examples of such genes which are dynamic just in the specific tissue of the erythroid cells of the blood, the silk organs of invertebrate organisms, the pancreatic islets of Langerhans, and the central tissues of the vertebrate eye, individually.

The results of luxury genes are in some cases alluded to as luxury proteins. The non-luxury genes, the supposed housekeeping genes, include sequences coding for histone or lactate dehydrogenase, which occur in all cells.

## Lyases

Group of proteins that catalyze double-stranded bond sequence or molecules.

## Lymph

Liquid found in the lymphatic system of vertebrates. Just likethe blood plasma of the blood it carries a significant number of the plasma proteins. A large portion of the cells in the lymphoid vessels is lymphocytes.

## Lymphatic System

Broad organization of vessels looking like veins (lymphatic vessels) emerging from little, thin-walled lymphatic vessels, which is presentin higher vertebrates. It isessentially involved in y recycling of interstitial liquid to the circulation system.

This is affected by dispersion of interstitial liquid through the lymphatic vessels to form lymph, and possible entry from the biggest lymphatic vessel, the thoracic pipe, into the left subclavian vein.

The lymphatic system likewise gives security against illness through lymph nodes blended all through the bigger lymphatic vessels, catching unfamiliar antigens before they enter the circulatory system.

## Lymph Node (Lymph Gland)

A glandular structure that is part of of lymphoid tissue found by and large around the vertebrate body, for example, in the axillae. The neck (tonsils and adenoids) and they are especially various in the peritoneal depression around the stomach (where they are known as Peyer's patches).

The lymph nodes are interconnected by vessels of the lymphatic framework and the different kinds of lymphocytes which make up a huge piece of the lymph node tissue can move from one node to another by means of the lymph. Lymph nodes are foci where attacking microbes and virus are in many cases concentrated by the immune system of the body, and irritation of the lymph nodes is a successive backup to bacterial or viral contamination.

## Lymphocyte

Circular cell involved in the immune system of vertebrates and bear a huge, round, and profoundly staining nucleus and very little cytoplasm. Little lymphocytes are incredibly heterogeneous in function regardless of comparative morphology.

## Lymphoid Tissue (Lymphatic Tissue)

Tissue of the lymphatic system with its lymph nodes and bigger collections of lymphoid cells. The organs that are made primarily out of such tissue include the spleen, thymus organ, lymph nodes, the bursa of Fabricius in birds, tonsils, adenoids and more modest nodes, for example, Peyer's patches near the digestive walls.

The addendum and bone marrow additionally contain significant quantities of lymphoid cells. Since the all-out number of lymphocytes in the human is around 2 x $10^{12}$ , in cell mass the lymphoid tissue is on a similar scale as liver and cerebrum. Lymphoid tissue is made out of different sorts of lymphocytes, which are gotten in fetal life from the liver and in grown-up life for the most part from the bone marrow.

From the marrow, numerous lymphocytes move out through the lymphatic vessels to colonize lymph nodes either for a brief time or forever, and the handling of the lymphocytes in this manner prompts the different populaces of T lymphocytes and B lymphocytes whose connections are so significant in the mechanics of the immune system.

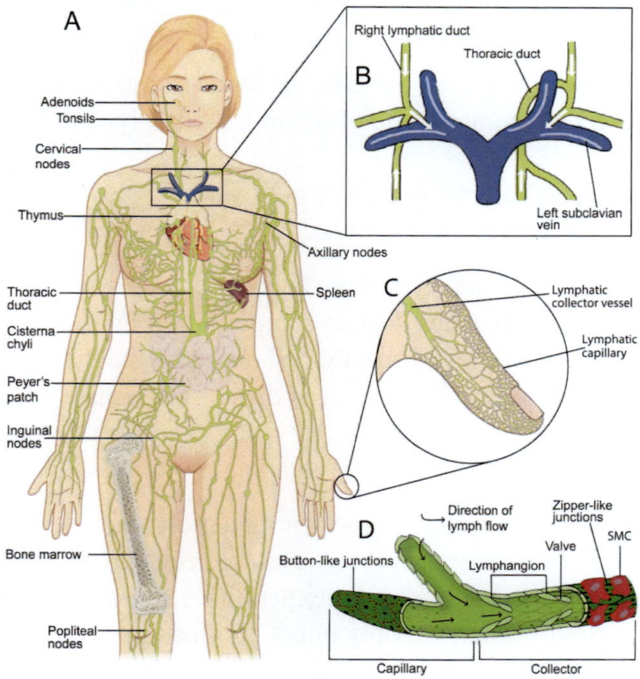


**Figure 28.** Anatomy Of The Lymphatic System.

*Source : Image By Wikimedia Commons*

## Lymphokines

Conventional name for a heterogeneous group of dissolvable, physiologically dynamic variables (m.=20000-80000), delivered by effector t lymphocytes when in touch with specific antigen. They are significant in cell-intervened invulnerability.

Lymphokines include macrophage inhibition factor (MIF), lymphocyte resistant factor (LRF), interleukins, y-interferon, specific macrophage arming factor (SMAF), chemotactic factors and cytotoxic elements.

## Lymphoma

Neoplastic development of the lymphatic tissue. One type of this cancer normal in central Africa is Burkitt's lymphoma; it includes monoclonal multiplication B lymphocytes which follows from disease with Epstein-Barr virus (most likely mosquito-borne).

A chromosomal movement close to the locus of the MGC oncogene on chromosome 8 is a continuous side effect in Burkitt's lymphoma.

## Lyon Hypothesis (Inactive X Hypothesis)

Proposed by Marylyon, recommending that irregular inactivation of everything except one of the X chromosomes in physical cells of the female is the system that accomplishes dosage compensation of X chromosomes in the homogametic sex of the warm-blooded animals. Inactivated duplicates of the x chromosome structure the Barr bodies of the female nucleus.

## Lyonization

Term in view of a translation of specific genetic genes expressed in the female as though she were conveying a x-connected recessive gene in a hemizygous condition (e.g., a lady who experiences hemophilia yet transmits this regularly recessive allele on only one x chromosome).

The peculiarity is depicted as lyonization with the understanding that one of the two x chromosomes is specially dense in most or every substantial tissue.

## Lysate

Result of cell lysis; frequently used to allude to a suspension of phage molecules delivered by the lysis of host microorganisms, or to a without cell homogenate following the activity of detergent or mechanical means to break the cells.

## Lysogenic Bacterium

Bacterium conveying a phage as a prophage. In this express, the condition of lysogeny, the bacterium develops regularly and no synthesis of phage occurs, yet affected by conditions, for example, bright light the condition of lysogeny can be broken, with ensuing phage increase and possible cell lysis of the host.

## Lysogenic Virus

Virus equipped for turning into a prophage by going into a steady relationship with a lysogenic bacterium. Lambda phage can occur in either a mild or harmful state in have microbes, and in the previous case it goes about as a lysogenic virus. It is desirable over allude to such infections as mild, in any case.

## Lysosomes

Intracellular vesicle encased by a single layer and comprising the primary site of intracellular digestive processes in eukaryotic cells. These bodies are especially rich in degradative compounds (e.g., phosphatase, proteases, lipases, nucleases, hydrolases) albeit some specialization of digestive properties are found with various lysosomes.

This large number of enzymes are usable at about pH 5, which is the kept up with pH of the lysosome. Material to be processed is normally bundled in an endocytotic vesicle (in the event that it has entered the cell from an external perspective) or in an autophagic vacuole (assuming it is of intracellular beginning).

Breakage of lysosomes in cell homogenates brings about the quick breakdown of numerous organelles and macromolecules, while cell death might result from the self-absorption of a cell by its own lysosomes.

## Lysozyme

Compound that digests glucose-amino-glycans and has a bacteriolytic activity. Tears and egg white are rich sources of lysozyme, and a few bacteriophages utilize the molecule to process the bacterial cell wall from inside after replication and subsequently work with discharge from the cell.

## Lytic Growth

Formative period of a harmful phage during which the viruses duplicate vegetatively and kill the host cells by lysis toward the end of the development cycle. Lytic development occurs additionally following enlistment of a lysogenic phage.

## Lytic Response

Lysis of microbes following their disease with harmful phage, rather than the lysogenic reaction where no lysis occurs.

## Lytic Virus

Virus that kills the host cells which it infects by causing cell lysis.

# M

## Ml3

Widely used cloning vector; single-stranded bacteriophage with a closed circular DNA genome. Since the product of replication of this phage, even when it carries an insert of some other gene, is a single-stranded DNA molecule, this implies that the molecules are readily used as a template for sequence analysis.

## Macroevolution

Evolutionary pathway or pattern viewed over a long geological time span. The evolutionary development of taxa above the species level are generally regarded as falling within this subject area.

## Macromere

Group of large cells that are produced in unequal cell divisions during early embryogenesis in such animals as sea urchins and mollusks.

## Macromolecule

Molecule with a molecular weight upwards of a few thousand Daltons. Most proteins and nucleic acids and many polysaccharides are macromolecules.

## Macronucleus (Mega-Nucleus)

Larger of two distinct sizes of nuclei occurring in ciliated protozoa. These nuclei are highly polyploid and are essentially responsible for vegetative growth, whereas the much smaller micronuclei (absent in the cells of some ciliate strains) have important sexual functions during conjugation.

## Macrophages

Group of several cell types that together form the reticuloendothelial system of metazoan organisms. Cells of this system have the common property of ingesting by phagocytosis and endocytosis molecules foreign to the organism such as bacteria, protozoa and cellular debris.

Macrophages are formed from circulating monocytes. A major feature in this transformation of cell types is an increase in the number of lysosomes.

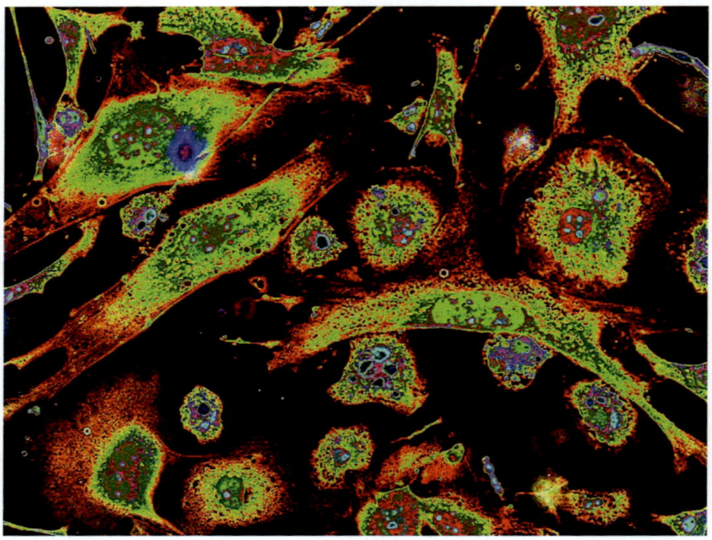

**Figure 29.** Epithelioid Cells And Macrophages In Peritoneal Cell Culture Of Mice.

*Source : Image By Wikimedia Commons*

## Main Band

In density gradient separation of genomic DNA, a broad peak containing most of the DNA and excluding any visible satellite DNA bands.

## Major Histocompatibility Complex (MHC)

Region of vertebrate chromosomes initially discovered through its control of immunological rejection of organ transplants and now known to mediate control of nearly all aspects of the immune response.

The genes making up the MHC are present on chromosomes 6 and 17 in man and mouse, respectively, and code for glycoproteins present on the surface of cells.

## Malaria

Disease caused by the single-celled parasite *Plasmodium falciparum*. Geographical areas of major incidence of the disease correspond closely with the distribution of sickle cell anemia. It is believed that the prevalence of the latter disease may be explained because individuals who are heterozygous for the sickle cell allele are more resistant to malaria and are therefore selected for in the population.

## Male Gametophyte

In higher plants, the pollen grain plus the pollen tube, containing three haploid nuclei.

## Malignancy

Attribute of cancerous neoplasms when they display characteristics of invasiveness and spread by metastasis. Neoplasms or tumors are not necessarily malignant, and when not, they are termed benign.

## Maloney Leukaemia Virus

Mouse retrovirus that produces lymphocytic leukemia. New born mice can be infected with the virus via the milk from an infected female mouse.

## Malpighian Layer (Stratum Germinativum)

Basal germinative layer of the vertebrate epidermis and composed of mitotically active cells and melanocytes. Sometimes taken to include the next most superficial layer-the stratum spinosum or spiny cells.

## Malpighian Tubule

Blind-ending tubular glands attached to the hind gut of insects and other arthropods, and serving an excretory function.

## Maltose

Disaccharide consisting of two glucose molecules joined by a glycosidic linkage. It is a degradation product of starch in germinating seeds such as barley.

## Mammary Tumor Virus

Retrovirus causing mammary tumors in mice and transmissible from mother to offspring via milk.

## Mannitol

Polyhydric alcohol derived from mannose or fructose. It is a white crystalline powder with a sweet taste and is widely distributed in plants and fungi.

## Maple Syrup Disease

Inborn error of metabolism affecting the degradation of valine, leucine and isoleucine by oxidative decarboxylation. Since degradation is blocked, blood and urine have markedly higher levels of these three amino acids and a corresponding increase in the αketo acids derived from them.

The deficiency gives the urine of affected patients an odor of maple syrup. Unless affected individuals are placed on a diet low in these amino acids early in life the disease normally proves fatal.

## Map Unit

Measure of the distance between the loci of two linked genes, that is two genes on the same chromosome. The map unit is generally regarded as being equivalent to 1 centimorgan (i.e., one-hundredth of a Morgan unit) which corresponds to a recombination frequency of 1 percent between the two genes.

## Marfan's Syndrome (Arachnodactyly)

Disorder of connective tissue in a human, inherited as an autosomal dominant. The disease is characterized by extremely long fingers and toes and has a prevalence of approximately one in 20000.

## Marker

Molecules of known size or identity that are run on electrophoretic gels to provide an indication of the expected position of other equivalent molecules of protein or nucleic acid. (2) molecules used to identify particular classes of lymphocytes by acting as specific antigenic conjugants. (3) gene, with a characteristic phenotypic expression and a known chromosomal location, used to provide a point of reference in determining the mapping position of other genes. (4) characteristic of a cell or cell line that enables separate cell identities to be established in mixed cell tissues.

Particular staining properties, or nuclear or nucleolar characteristics, or even an aberrant mitotic chromosome, may serve to identify the cells of a cell line. Thus, if marked quail cells are injected or transplanted into a chick embryo, the destiny of the implanted or injected cells may be followed in tissue sections if the quail cells carry a marker such as an enlarged nucleolus.

## Marker Rescue

Peculiarity in which a populace of microorganisms is infected with two unmistakable markeeed phages (for example Two sorts that can be genetically recognized), yet following the irradiation of one of the two phage strains. In view of recombination, a portion of the prophage come to have properties of the irradiated strain.

## Masked mRNA

Messenger RNA present in a cell, however not being effectively converted into protein. This is related to specific sorts of proteins as messenger ribonucleoprotein (mRNP) molecules and can be initiated to work in protein combination by unambiguous physiological triggers (e.g., molecules).

## Mast Cell

Cell type found in connective tissue and described by basophilic cytoplasm and capacity of mixtures like heparin and receptor. During inflammatory reactions, these substances might be set free from macrophages as secretory granules.

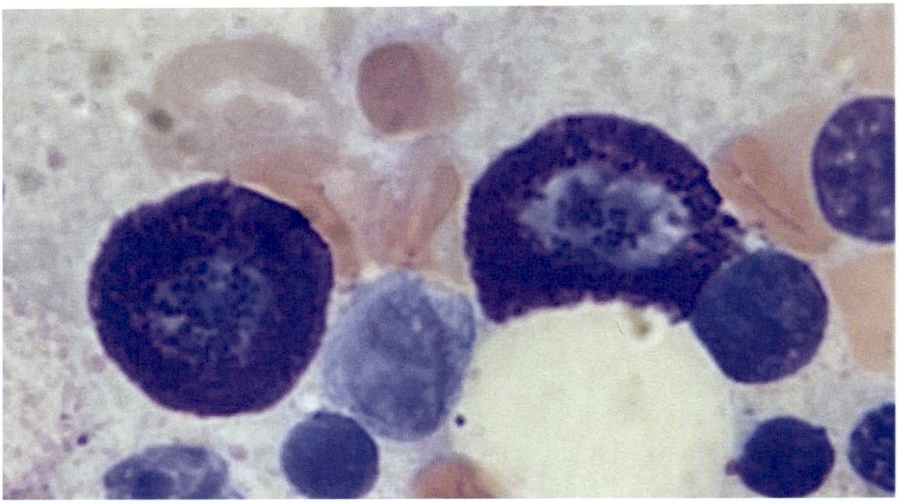

**Figure 30.** Two Mast Cells In Bone Marrow At 100x Magnification After Wright Staining.

*Source : Image By Wikimedia Commons*

## Master Gene

Gene that controls different genes, especially when the genes controlled are totally expressed naturally in a specific sort of separated cell.

## Master Slave Hypothesis

The hypothesis made by H.G. Callan that most eukaryotic genes existed in different indistinguishable duplicates (the slaves) and that the successions were completely kept indistinguishable by continuous adjustment, utilizing a 'solitary expert' duplicate of the gene as a layout for amendment of the 'slave' duplicates.

Aside from conceivable rectification components in ribosomal gene duplicates, the speculation has not been validated by proof.

## Mastigonemes

Short hairs that project horizontally from a few eukaryotic flagella, typically in a couple of columns. Some mastigonemes are moderately solid and hollow, others are adaptable however strong, the two kinds occuring in different organisms.

## Mate Killer

CA cell of *Paramecium aurelia* that kills a non-killer (sensitive) cell with which it forms. The killing property gets from the presence of mu-particles inside the cell.

## Maternal Effects

Impacts of the aggregate ~r genotype of the mother on the progeny, for the most part of a temporary nature. Subsequently maternally received protein or messenger RNA put away in the egg prompts maternal impacts in a few early incipient organisms.

## Maternal Legacy

Genetic impacts that are expressed by means of the maternal line. There are many such impacts, some subsequent from the absolutely maternal beginning of mitochondria or chloroplasts, others from consequences for the aggregate of the undeveloped organism by early improvement inside the female, like the maternal effect on snaking in snails.

## Mating

Sexual multiplication between sets of life forms of two particular sexual types.

## Mating System

Example of multiplication inside an animal types, in some cases including sex proportions contrasting from 1:1, some of the time including changes in individual sex with time or situation, some of the time including absolute parthenogenesis and complete shortfall of mating for part or the entirety of the yearly cycle.

## Mating Type

Subpopulation of a microorganism that is assigned as having specific mating genes. There might be multiple particular mating types (as in certain organisms) and a complicated sequence of inter-compatibility. Integral mating types frequently remember each other by specific cell surface receptors.

## Mating Type Gene

Grouping that controls the decision between two mating types known as an and a; most popular in the yeast Saccharomyces. Haploid yeast cells every so often switch their mating type.

At a genetic level such a switch has been displayed to include extraction of a specific mating type gene grouping from the genome, with its substitution by a gene of the elective mating type. This cycle has come to be alluded to as a tape component.

## Matrix

In cell science, a free meshwork of molecules, particularly the extracellular system, formed by many tissue cells. This complex of proteins and polysaccharides is particularly significant in connective tissue which is the fundamental constituent of skin and bone.

Cell walls of microorganisms, growths and plants are examples of extracellular systems. (2) channel or other retentive material utilized as a clean surface or three-layered region for cell development in tissue culture.

## Matrocliny

Circumstance in which progeny look like the female parent more intently than the male parent. In Drosophila, a model is managed by females of appended x females.

## Maturase

Protein that catalyzes the removal of introns. Almost certainly, such proteins are not enzymes, however, go about as elements that adjust the movement of the grafting catalysts. Numerous maturases are themselves results of the sequences whose right joining they help to control, and in any event some are coded on introns (e.g., maturases found in yeast mitochondria).

## Maturation Division

A Molecular division during which the chromosomal supplement of a nucleus is changed in accordance with some ideal sum. In this manner, the divisions associated with meiotic divisions to deliver haploid gametes are development divisions.

## Maxi-cell

Cell of *E. coli* or different microorganisms that has been UV-illuminated, yet at the same time holds some genomic DNA.

## Maximum Permissible Dose

Most extreme degree of ionizing radiation that current security norms license over a unit period.

## Mcardle Illness

Innate illness in humans because of a lack of glycogen phosphorylase in skeletal muscle, the gene for which is encoded on chromosome 11. The sickness includes glycogen capacity and has an occurrence of around one out of 500000.

## Megakaryocyte

Amoeboid giant cell found in the bone marrow. It contains a huge, multinucleolate, polymorphous nucleus. Gotten from haemocytoblasts, megakaryocytes presumably lead to blood platelets, the development of which is found in the megakaryocyte cytoplasm.

## Megaphyll (Macrophyll)

One of two sorts of leaves found in pteridophytes (which includes plants like ferns). Leaves of angiosperms and gymnosperms are at times additionally assigned as megaphylles.

## Megaspore

One of the four results of the meiotic division of the megasporocyte in angiosperm plants. Three out of the four haploid megaspores don't mature, while the fourth divides to form the female undeveloped organism sac or gametophyte.

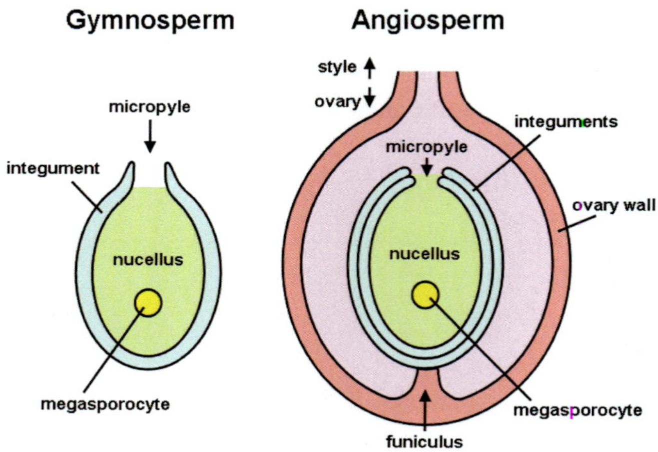


**Figure 31.** Ovules Diagram.

*Source : Image By Flickr*

## Megasporocyte

Megaspore mother cell of angiosperm plants. Meiotic division of this cell brings about the megaspores, one of which develops into the incipient organism sac.

## Meiosis

Interaction of cell division engaged with the synthesis of the male and female gametes from their diploid begetter cells. Since the course of preparation brings about the synthesis of a combination nucleus with chromosomes of the two gametes, it is important to lessen the chromosome supplement of the gametes to a haploid number so treatment delivers a diploid as opposed to tetraploid cell.

Various parts of meiosis are specific and very particular from mitotic division. They can be set out as follows.

1) prophase is significantly more drawn out in meiosis than in mitosis and is partitioned into five continuous stages: leptotene, zygotene, pachytene, diplotene and diakinesis (additionally called as leptonema, zygonema, pachynema, diplonema and diakinesis, separately). 2) in certain living beings, including vertebrates, the prophase is delayed to the point that the chromosomes are

transcriptionally active, creating the trademark lamp-brush chromosomes of amphibian diplotene stages.

3) homologous chromosomes partner together to form a quadruplicate of four chromomeres; homologous chromosomes don't relate at mitosis.
4) genetic recombination occurs during meiosis and the sequence of chiasma during the time spent getting over gives visual proof for it. This includes non-sister chromatids of the chromosome quadruplicates.
5) DNA replication goes before the leptotene phase of the first prophase, however some extra DNA replication (about percent of that expected for full replication) occurs during diplotene to join together the sequences that have become isolated by crossing-over. This outcomes in the sequence of intermediates. No DNA replication occurs between the two meiotic division cycles.
6) two progressive molecular divisions occur during meiosis, yet only one during mitosis. The chromosomes don't significantly uncoil between the two meiotic divisions.

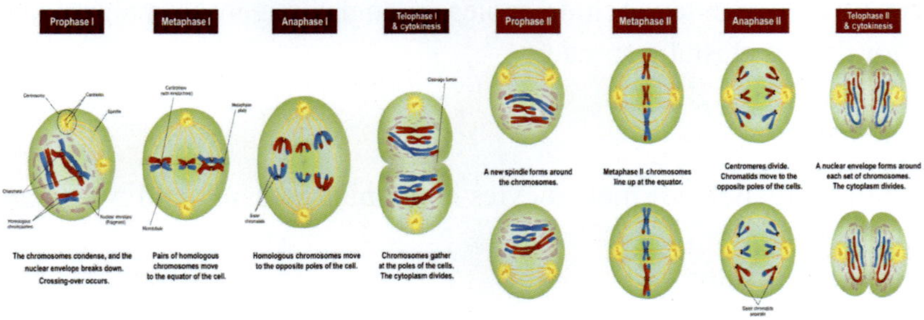


**Figure 32.** Meiosis Stages.

*Source : Image By Wikimedia Commons*

## Meiotic Drive

Any of various systems that produce inconsistent quantities of the two gametic types formed by a heterozygote.

## Melanic

Portraying a person that, due to expanded melanin synthesis, is a lot darker than typical. Melanism is normal in birds, well evolved organisms and once in a while occurs in insects and different invertebrate organisms.

## Melanin

Dark or earthy colored pigment that creates the pigmented shade of skin, hair and the retina. It is contained in melanophores and comprises of a complex of protein and tyrosine subordinates.

## Melanocyte-Stimulating Hormone (MSH, Intermedin)

Polypeptide molecule, delivered in the moderate projection of the pituitary organ, that causes the spread of melanin in melanophores and in this way an obvious coloring of the skin.

## Melanoma

Cancer brought about by increased development of melanocytes. Harmless cancers of melanocytes include moles, though dangerous melanomas are noted for their metastatic spread.

## Melanosome

Intracellular organelle of melanocytes containing melanin and tyrosinase.

## Melanotic Cancer

Cancer that contains melanin. One melanotic growth in Drosophila comprises of a collection of darkened cells at different formative stages and results from melanization of blood cells.

## Melatonin

Engineered result of serotonin discharged by the vertebrate pineal organ. It prompts coagulation of melanin in dermal melanophores and in this way easing up of the skin tone.

## Membrane Potential

Difference in potential across the cell membrane The membrane potential varies in magnitude amongst cells of different kinds, ranging from -20 to

-100 my. A differential distribution of charged molecules (e.g., ions) across the cell membrane generates the membrane potential, which is traditionally expressed as negative to show that the inside of the membrane is negative with respect to the outside. Active (energy-demanding) processes and the membrane's differential passive (non-energy-demanding) permeability to distinct ions produce this ion distribution (concentration gradient).

Changes in membrane permeability result in a shift in ion distribution and, as a result, a shift in membrane potential. The conduction of a nerve impulse in neurons is based on a local drop in membrane potential or depolarization of the membrane propagated along the axon.

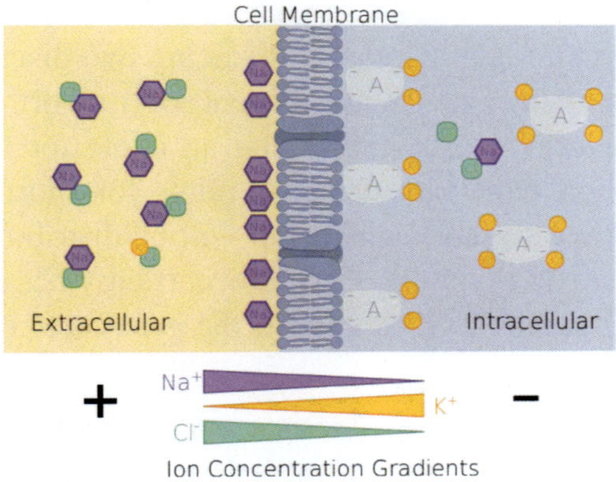


**Figure 33.** A picture of Membrane Potential.

*Source: Image by Wikimedia Commons*

## Membrane Transport

Biological membrane transport as most polar compounds are fundamentally impermeable to the lipid bilayer of the membrane. Cells have evolved specific mechanisms for transporting molecules across their membranes. These include membrane transport proteins that act as channel proteins and carrier proteins, the latter of which is engaged in facilitated diffusion. Membrane transport also include sodium pumps.

## Meme

Term coined by R. Dawkins to describe a pattern of information or transmissible notion that can transfer from brain to brain, computer to computer, and therefore acquires transmissible element properties.

## Memory Cell

Immune cell that stores immunological memory, allowing an organism to produce a secondary immune response when confronted with an antigen for the second time.

## Mendelism

The inheritance of chromosomal genes according to the laws postulated by Gregor Mendel.

## Mendel's Laws

Independent assortment law: Factors influencing two discrete genetic traits (now identified as distinct genes at different loci) assort separately. Thus, a person can have brown hair, expressing an allele for brown hair color at one site, and short hair, expressing an allele for short hair length at a different locus. Hair color and hair length will be distributed differently in the offspring of such an individual. The law currently has a caveat: the two loci must be on different chromosomes; if the genes are connected, the law does not apply.

Law of segregation: Gametogenesis separates the factors that influence a character. In current terms, this indicates that each gamete has one of each pair of alleles, whether homozygous or heterozygous.

## Mericlinal Chimaera

Organism or component of an organism in which two genetically separate cell lines exist, one of which is partially inside the other. Plants with chimera produce two specific types, if not two distinct species, when they grow together.

## Meristem

Mitotic activity is high in this plant tissue, which gives rise to cells that make up the plant's tissue. There is a distinction in active meristems between those cells that stay meristematic (the starting cells) and those that differentiate. The primary plant body is formed by primary meristems (e.g., apical meristems) present in the shoot and root tips.

Some apical meristems have been shown to have the ability to maintain mitotic activity indefinitely. The nucleus to cytoplasm ratio in these early meristem cells is typically high. Plastids are found as proplastids in the cytoplasm, which is normally thick with little vacuolation. The meristem also produces leaf primordia in the shoot tip. These develop into peg-like projections, after which numerous meristems working concurrently or sequentially contribute to the leaf's growth.

The root cap receives cells from the distal meristem, whereas the rest of the root receives cells from the proximal meristem. The quiescent centre separates these two meristems. Axillary meristems form in the leaf axils, although the apical bud's influence often suppresses division activity in these meristems. Many monocotyledons have intercalary meristems that are found away from the apical meristem, such as in the nodes and leaf sheaths. Secondary meristems (for example, the cambium and phellogen) are lateral meristems that give rise to the plant's secondary body.

## Meristematic

Referring to meristems, which are undifferentiated plant tissue capable of repetitive mitotic division.

## Meroblastic Cleavage

The cytoplasm of the egg is divided, resulting in an unequal distribution of yolk between the resultant cells, resulting in an embryo with cells of varying sizes.

## Merogony

When a part of an egg develops into a full but small embryo. Microsurgery is used to create egg fragments that are normally nucleated (albeit the nucleus may be haploid) but have a reduced amount of egg cytoplasm.

## Meromyxis

Bacterial conjugation which is a process in which only a few genes are transferred rather than the full genome.

## Merozygote

It's a partially diploid bacteria formed when a genome fragment from a donor is transferred to an entire recipient cell. Bacterial conjugation, transduction, or transformation can all be used to transfer information.

## Meselson and Radding Model

Advancement to the holliday model of chromosome recombination. The twisting and isomerization of strands in the cross-over intermediate is presented as a way to overcome the nicking requirements of the original holliday model.

## Meselson and Stahl Experiment

This was an experiment in molecular genetics which was first published in 1958. It demonstrates that DNA replication is semi-conservative. The procedure entailed labeling *E. coli* DNA with 15N and then moving the bacterium to a medium that only contained 14N. Cesium chloride density centrifugation was used to follow the dispersion of the DNA that had integrated the two isotopes, and the presence of a band partially labeled and hence of intermediate density provided evidence for a semiconservative pattern of DNA replication.

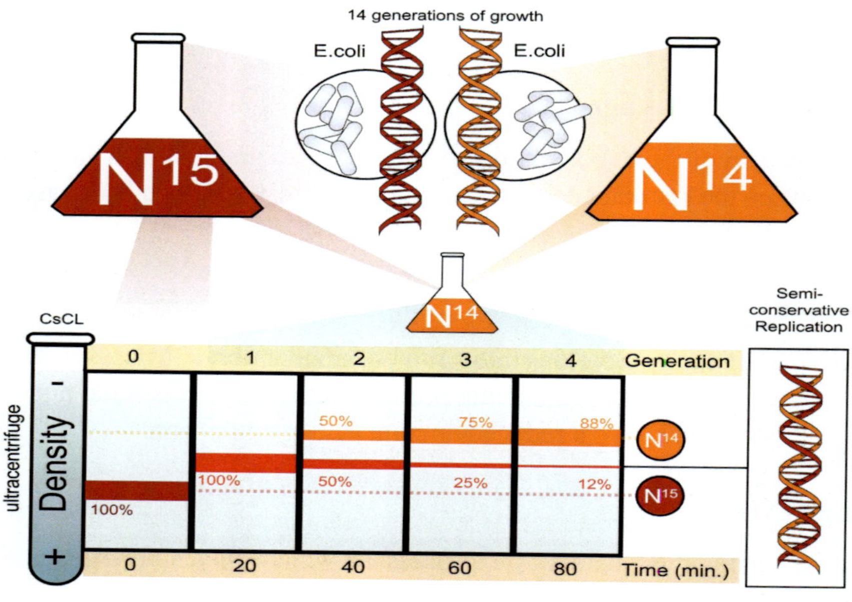


**Figure 34.** A diagram of Meselson and Stahl experiment.

*Source: Image by Wikimedia Commons*

## Mesenchyme

Type of connective tissue. It is abundant in embryos and consists of amoeboid cells with many pseudopods. Mesenchyme is primarily a derivative of mesoderm, and it gives rise to connective tissue, erythroid and lymphoid pathway cells, and cartilage during development. Between mesenchymal and epithelial tissue, there are many critical embryonic tissue interactions.

## Mesoderm

Refers to the middle of the three cell layers of triploblastic animal embryos. The mesoderm grows in two layers in coelomates (animals having a body cavity): an inner splanchnic mesoderm and an outer somatic mesoderm. The former develops into visceral organs in vertebrates, whereas the latter develops into somites, which create muscular and skeletal parts. The mesoderm is the progenitor tissue for blood and kidneys in general. Gonads. Nodes of lymph, cartilage, bone, and muscle

## Mesoglea

Literally middle jelly; Coelenterata has a gelatinous layer between the ectoderm and the endoderm (e.g., hydra and jellyfishes). It does contain cells, but they are most likely to have migrated from the animal's inner or outer cell layers.

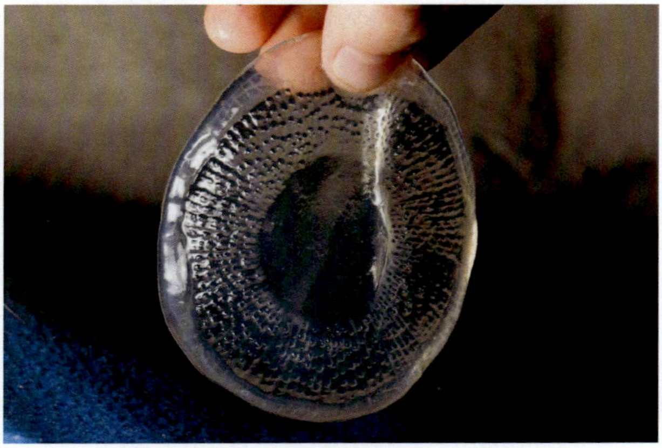

**Figure 35.** The mesoglea from a Lion's Mane Jelly.

*Source: Image by Flickr*

## Mesothelium

Certain forms of epithelium generated embryonically from mesoderm or mesoblast; used in the lining of coelomic cavities and the synovium of diarthrodial joints, in particular.

## Metabolism

All living organisms have a highly interconnected network of enzyme-catalyzed processes. Although the word refers to all such processes, it is frequently used to refer to the reactions involved in the degradation of a specific molecule.

## Microfilament

Polymerized actin intracellular Ffber with a diameter of 5 to 7 nm. Many cellular extensions, such as the microvillus brush border of many intestinal cells and the hair cells of the cochlea of the ear, include nuclei of cross-linked actin filaments. Many cells have dynamic actin filament structures, which appear to operate as sensory devices, allowing the cell to detect specific elements of its immediate environment.

Oriented microfilament arrays are not limited to cell surface protrusions, as all eukaryotic cells include additional bundles of these filaments going through the cytoplasm. A belt-like contractile ring of microfilaments, for example, helps to contract the nucleus part of the cell right below the plasma membrane during cell division.

Microfilaments make up belt desmosomes, which are seen in many epithelial cells, as well as stress fibers. Adhesion plaques are formed when actin filaments are attached to cell membranes at least on one end. The exact process of this adhesion remains unknown.

## Microinjection of Cells

Direct injection of molecules (such as antibodies, mRNA, and DNA) or organelles (such as nuclei and mitochondria) into cells using ultrafine glass needles. Although reasonably large tissue culture cells can be injected, eukaryotic animal eggs are preferred since they are often quite large; direct injection into the cell nucleus is achievable in some circumstances. Some of the DNA injected into the cell's nuclear genome may become permanently included. If the cell is a fertilized egg, the animal and its offspring may undergo genetic change.

## Microspectrophotometer

Light microscopy of cells or tissues is combined with spectrophotometry, an electronic method of detecting the absorbance of light of any wavelength. This apparatus can be used to quantify dye binding or ultraviolet-absorbing material concentrations in transects across a cell or in a specific cytoplasmic or nuclear location.

## Midwife Toad

Amphibians get their name from the male's behavior of carrying fertilized eggs and young tadpoles on his back and linked to his hind legs. Kammerer stated in 1923 that he had conducted trials in which usually terrestrial animals were housed in water and bred from such aquatically adapted stock. In the aquatic adaptation, male toads gained a modification in the form of their foot pads, which they transmitted on to their offspring. These tests, which support Lamarckism, have never been successfully replicated, and are today regarded to be defective in execution or interpretation.

## Mimicry

A resemblance in appearance between two different species that protects one or both from predation. Batesian mimicry occurs when an organism that is known to be harmful or toxic is made to seem like another creature that is known to be dangerous or poisonous. Both species gain from positive reinforcement of the link between color and disagreeable taste in Mullerian mimicry since both species are unpalatable.

## Mitochondrial DNA (MtDNA)

It refers to the DNA of mitochondria. It usually consists of a closed circle of double-stranded DNA with a length of between 5 and 30 #lm, a size of 15-20 kilobases in most animals (greater in high plants), and approximately 50 gene sequences, about half of which code for transfer RNA. Because the DNA is not conjugated with histone, it does not exist as nucleosomes. The MtDNA of some organisms, including the *Drosophilia melanogaster*, has been completely sequenced.

## Mitogen

Refers to the substance which stimulates mitosis of cells. Mitogens, many of which are derived from plants, are particularly relevant in immunology because they can cause polyclonal lymphocyte populations to divide and change without being directly stimulated. Mouse T lymphocytes are stimulated by both concanavalin A and phytohaemagglutinin, whereas mouse B and T lymphocytes are stimulated by pokeweed mitogen. Mitogenic factors are lymphokines that can cause lymphocytes to multiply.

## Mitosis (Karyokinesis)

Nuclear division in eukaryotic cells, but not the same as that involved in gamete formation. In bacteria, mitosis does not occur, and in certain lesser organisms (e.g., Physarum), the nuclear membrane does not break down during mitotic division. Mitosis ensures that each daughter cell has a complete diploid set of genes and chromosomes after division.

DNA replication does not occurduring mitosis; it occurs during the s phase of the cell cycle and is completed before nuclear division begins. Because gene transcription is temporarily halted during mitosis, RNA synthesis is also halted. Mitosis begins with chromatin condensation (prophase), with the chromosomes forming lengthy double strands. Each is made up of two chromatids, which are the result of DNA replication during the s phase.

The nuclear envelope breaks down near the end of prophase, the nucleolus vanishes, and a nuclear spindle of microtubules emerges. This spindle is made up of centrioles in animals and lower plants, and each of its poles has a centrosome, which is a microtubule aggregation area. The centromeres of the chromosomes attach to specific spindle fibers, and the chromosomes align in a plate-like pattern across the equatorial plane of the spindle during the following stage of mitosis, metaphase.

In contrast to meiosis, the two homologous copies of each chromosome do not connect during mitosis. During the next phase, anaphase, the distinct centromeres of each chromosome's twin chromatids are pulled towards the poles of the spindle by their spindle attachment, pulling the chromatids apart. When each chromatid has become a different chromosome, and each daughter set of chromosomes is distinct and independent, this is accomplished at telophase.

At this stage, the chromosomes also lengthen and become less condensed. The two new nuclear envelopes develop around the two chromosomal clusters at the end of telophase, nucleoli return, and the cytoplasm splits to produce the two new daughter cells.

## Modification Enzymes

In bacterial cells, methylases attach methyl groups to DNA at certain places, particularly adenine bases. Methylation prevents genomic DNA from being digested by its own restriction endonucleases, which won't cut methylated nucleotides. The restriction enzymes defend the bacterium from invasion by DNA phages, whose DNA is not methylated and hence susceptible to destruction.

## Monoclonal Antibody

Individual clonal populations of hybrid cells- hybridomas- generate and secrete antibodies after fusing individual B lymphocytes from an immunized animal with individual lymphocytic tumor cells. The hybridoma cell is immortalized by the cancer cell, while the B cell continues to release its single particular antibody type. Hybridomas containing each chosen clone can thus be produced and grown in large quantities to obtain copious amounts of specific and homogeneous antibodies from each.

Immunizing experiment animals with an antigen, on the other hand, produces a "cocktail" of various antibodies, each made by a different subpopulation of B lymphocytes and each directed against a different epitope on the original molecule. Polyclonal antigen preparations can generate cross-reactivity due to epitopes shared with different antigens or activity against small impurities in the original antigen preparation. The development of monoclonal antibody technology has solved this difficulty and greatly expanded the technical and diagnostic repertoires of cell biologists and pathologists.

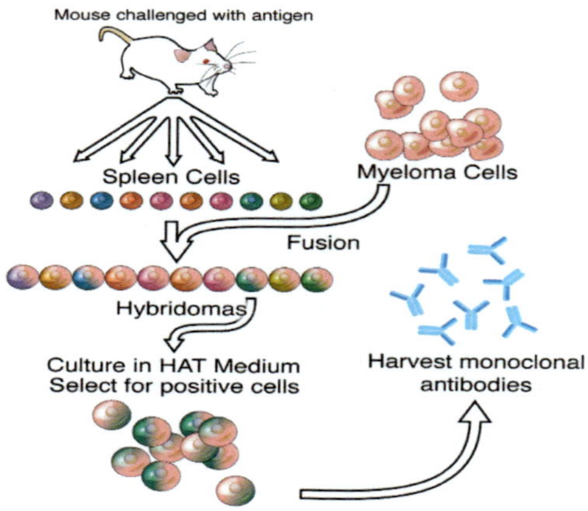


**Figure 36.** A monoclonal antibody diagrams.

*Source: Image by Wikimedia Commons*

## Monocyte

Monocyte refers to the large agranulocytic leucocyte blood cell of vertebrate blood. Monocytes, which make up around 7% of all human white blood cells, are 12-20μm in diameter and have oval nuclei. They are phagocytic and capable of fast migration to other tissues during inflammation, especially in regard to bacteria and big particulate debris. They are known as macrophages when they are in the tissue.

## Morphogenesis

Form and structure development of a tissue or organism. Although cell differentiation is frequently involved in morphogenesis, several aspects of shape formation are clearly autonomous. A population of cells may be differentiated in a specific fashion, but due to a morphogenesis mutation, they do not collectively adopt the right form of the tissue or organism in which they are found.

## Morula

Embryos formed through holoblastic cleavage; a stage attained by (cleavage when the embryo consists of a solid spherical mass of blastomeres, usually without a cavity, which would make it a blastula if one existed.

## Mucosa (Mucous Membrane)

Many vertebrate cavities in contact with the external environment have a layer of moist surface epithelium. Much of the respiratory, urinogenital, and digestive tracts are included. The mucus that coats the mucosa is high in mucoproteins and glycoproteins, particularly mucin, and is released by goblet cells located in intestinal villi, for example. The mucus is secreted elsewhere because much of the mucosa of the urinogenital tract, for example, lacks goblet cells.

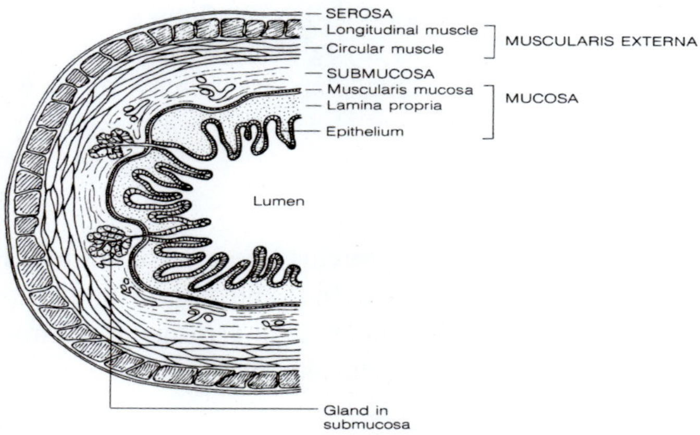


**Figure 37.** A labeled diagram of Mucosa.

*Source: Image by Wikimedia Commons*

## Mutagen

A mutagen is a chemical or physical substance that causes more mutations than natural mutations. Ionizing radiation, UV light, and chemicals are all important mutagens (e.g., nitrogen mustard and nitrous acid). The mechanisms at work are diverse: By covalent interaction between neighboring pyrimidine bases on the DNA chain, ultraviolet light stimulates pyrimidine dimer synthesis; nitrous acid removes an amino group from cytosine, turning it to uracil.

## Myeloid Tissue

In the fetal liver and spleen, as well as the red marrow of certain adult bones, vertebrate erythrocyte and leucocyte-producing tissue is discovered which is referred to as myeloid tissue.

## Myofibrils

Myofibrils are banded, rod-like assemblages of myofilaments that are present within the sarcoplasm of striated muscle cells (muscle 'fibers') and are responsible for contraction. Actin, alpha-actinin, and myosin are arranged in highly organized, interdigitating arrays in myofibrils, giving them their typical banded appearance.

## Myoglobin Gene

The myoglobin gene codes for the protein myoglobin. Unlike globin genes, which have two introns apiece, the myoglobin gene has three introns, each of which is so long that the exons make up less than 5% of the overall gene coding sequence.

## Myosin

One of the most important muscle proteins (the other is actin). The myosin molecule has a head (heavy meromyosin) and a tail (light meromyosin); the globular head has sites for attaching actin and the tail permits filament formation by aggregating with other myosin molecules.

# N

## Naegleria

Naegleria is a genus of soil amoebae best known for its transition from amoeboid to flagellated cell form under experimental settings involving variations in the ionic content of the fluid in which the cells are suspended.

## Nail Patella Syndrome

Human inherited disease caused by a dominant gene on chromosome 9 that manifests as undersized or nonexistent knee - caps and malformed fingernails.

## Natural Acquired Immunity

Natural acquired immunity is the ability to resist disease after being exposed to the disease-causing virus or organism at some point in the past.

## Natural Immunity (Native Immunity, Nonspecific Immunity)

Natural immunity refers to mechanisms for neutralizing and removing foreign organisms or non-self-molecules from animals that do not involve an immune response, including specific antigen recognition, lymphocyte activation, antibody synthesis or initiation of cell-mediated immune phenomena, and memory for the evoking antigen.

Natural immunity includes the production of interferon by virally infected cells, the action of lysozyme, the ingestion of harmful agents by phagocytes, the encapsulation of foreign bodies, the cytotoxic activity of NK cells against virally infected or neoplastic cells, and the action of various humoral lysins, agglutinins, and precipitins. Some natural immunity is a product of the host's genetic makeup (e.g., human immunity to canine distemper). Natural immunity is critical in invertebrates for defense against hazardous chemicals, and it complements the immune system in vertebrates.

## Natural Selection

Selection of individuals displaying greater fitness in terms of fertility and fecundity as opposed to other individuals within the same population is a mechanism engaged in evolution. The rise in population fitness over time is proportional to the genetic diversity in fitness among its individuals. Given some degree of population control by variables such as predation, sickness, and limited food availability, the progeny of superior individuals is preferentially selected, resulting in an increased representation of their parents' genes in the gene pool of the population.

## Nature and Nurture

The independent influences of heredity and environment on an organism are referred to as its expression. Even at the molecular level, the appearance of an organism - the phenotype - is the product of an interaction between genotype and environment, and the expression of all genes is influenced by environmental factors in some way. When an illness or phenotypic characteristic is described as genetically determined, it should be understood to mean that the genetic component is very strong, whereas the environmental component is minor or non-existent.

## Nearest-Neighbor Analysis

Technique for determining the relative frequencies with which pairings of each of the four bases appear in neighboring places in DNA sequencing. Even if DNA molecules from different origins have the same base content, their nearest-neighbor analysis can be very different.

## Nebenkern

Nebenkern is the medial section of a spermatozoon. In this, there is a helical arrangement of clumped mitochondria which in turn occurs at the proximal end of the tail filament.

## Negative Complementation

When allele complementation allows a mutant component in a multimeric protein to inactivate the wild-type subunit with which it interacts.

## Negative Control

Regulatory gene expression by attaching a certain controlling element to DNA and preventing transcription as a result. Negative gene expression control is most likely the most important mechanism in both eukaryotic and prokaryotic gene expression. The lac operon in *E. coli* is negatively controlled, meaning that transcription will occurif the repressor protein that prevents operon activity is absent. Thus, inducing the operon is a form of de-repression.

## Negative Interference

The tendency for one recombinational event and the resulting cross-over to boost the chances of another comparable exchange occurring nearby.

## Negative staining

Staining technique for viruses and other particulate molecules in electron microscopy. An atomizer sprays a suspension of the substance to be investigated with a phosphotungstic acid solution onto a carbon-coated grid. This approach makes the minute details of the specimen's contours more visible.

## Negative Supercoiling

A duplex DNA molecule supercoils on itself in the opposite direction of the double helix strands' rotations.

## Neomorph

The wild-type allele produces a fundamentally novel and distinct effect than the mutant allele.

## Neomycin

Aminoglycoside antibiotic generated by various *Streptomyces* species that has a broad spectrum of antibacterial activity. It is unsuitable for systemic use due to toxic adverse effects. The biochemical effect of neomycin is to block protein synthesis on the 70S (prokaryotic) ribosome; it is similar to streptomycin in this regard.

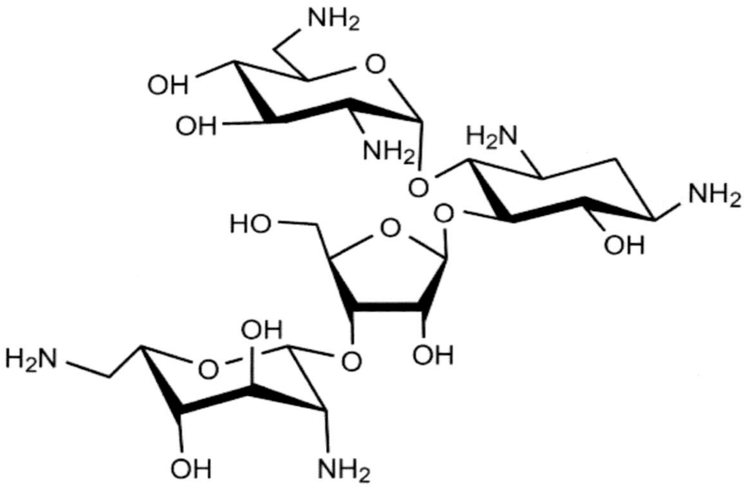


**Figure 38.** Structure of Neomycin B with correct stereochemistry.

*Source: Image by Wikimedia Commons*

## Neoplasm (Tumor)

A population of cells that have proliferated at an abnormal pace and are thus resistant to the factors that ordinarily govern the organism's growth and cell division. A neoplasm can be malignant, and hence cancerous, if it has invasive growth characteristics and spreads through metastases, or benign if none of these features are present. A benign neoplasm is a common wart caused by HPV, whereas a malignant neoplasm is a carcinoma.

## Neoplastic

A cell with irregular cell proliferation, such as that found in tumors. Neoplasms can be benign and non-invasive, like a common wart, or they can be malignant, like cancer.

## Neoteny

The retention of larval or juvenile characteristics in an animal after the usual juvenile stage has passed. Neoteny is hypothesized to have had a role in evolution, such as in the evolution of vertebrates from an ascidian tadpole-like organism that repressed metamorphosis to the sessile adult form.

**Figure 39.** An Image of Neoteny.

*Source: Image by Wikimedia Commons*

## Neural Crest

The neural plate margin/tips of the neural folds form a population of cells. The cells travel widely throughout the vertebrate embryo's body, giving rise to a variety of tissue types. Peripheral neurons, the autonomic nervous system, some glia, some cartilage, bone and connective tissues of the skull, odontoblasts, the adrenal medulla, and pigment cell lineages are all examples of derivatives. Its phylogenetic appearance was almost certainly a need for vertebrate evolution.

## Neural Fold

With the exception of teleosts, all vertebrates construct the majority of the CNS by rolling up the neurectoderm into a tube oriented longitudinally along the embryo's anterior/posterior axis. The synthesis of two bilateral neural folds - elevations of the neural plate's edge - which then proceed to travel towards the midline, results in rolling morphogenesis. Fusion occurs at the point of contact, resulting in a longitudinally oriented neural tube coated in the ectoderm. This is known as neurulation, and the embryo is known as a neurula at this stage.

## Neural Plate

The underlying chordamesoderm causes ectoderm to become neurectoderm. The word usually refers to the period during neural tube synthesis when the neurectoderm cells have become detectably columnar, generating an epithelium thicker than the surrounding ectoderm, but before the elevation of the neural folds.

## Neural Tube

Tubular primordium is the embryonic predecessor to the vertebrate CNS (produced during neurulation); it develops anteriorly into the brain and posteriorly into the spinal cord. (In teleosts and at the sacral extremity of other vertebrate embryos, the neural tube is created by cavitation of a solid mass of neurectoderm cells, rather than elevation and fusion of neural folds.)

## Neuraminic Acid

Many phospholipid membranes contain this substance. It's a glycccoprotein that's particularly common in neuron plasma membranes.

## Neuraminidase

Removal of sialic acid (n-acetylneuraminic acid) residues from the outer surfaces of plasma membranes by an enzyme.

## Neuroblast

Nerve cells are produced by the division and differentiation of embryonic cells.

## Neurofibromatosis

An autosomal dominant gene with full penetrance causes a single disease in humans. Numerous cutaneous neurofibromatoma tumors characterized it. Mental retardation and spinal cord compression are sometimes present, as well as coffee-colored patches (cafe-aulait patches).

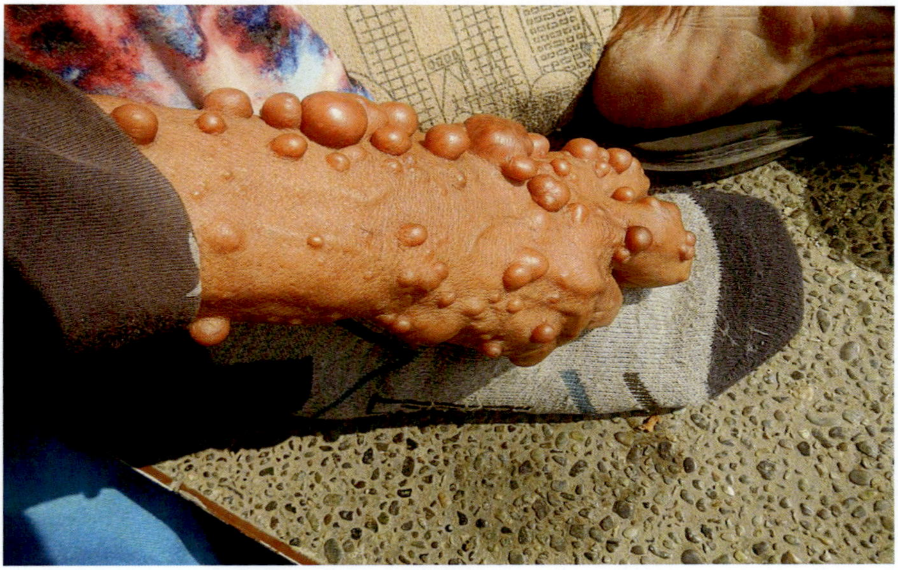

**Figure 40.** Individuals Suffering with Neurofibromatosis.

*Source: Image by Wikimedia Commons*

## Neurofilament

Filament found within nerve cells that is made up of three separate intermediate filament polypeptides and is thought to help axons strengthen.

## Neurohormones

Animal nervous system neurosecretory nerve cells produce a group of molecules that have effects on tissues far from the site of synthesis. The majority of neurohormones are peptides. Vasopressin, oxytocin, and noradrenaline are among them. Neurohormones are typically synthesized in the neuron's cell body and transmitted down the axon to the nerve terminal. Although some have targets near the nerve ending, the majority of them enter the bloodstream before reaching their intended region. The hypothalamus of the brain is where these hormones are mostly synthesized in birds and mammals.

## Neurological Mutant

Organisms with aberrant movement or behavior that can be traced back to a central nervous system abnormality or malfunction.

## Neuromuscular Junction (Myoneural Junction)

The connection between a nerve's motor endplate and the muscle cell it activates. When an action potential is received at the endplate, a chemical transmitter (neurotransmitter) is released and diffuses across the space between nerve and muscle cell membranes. The transmitter subsequently binds to receptors on the cell membrane of the muscle cell, causing it to depolarize. This sets off a chain of processes that result in muscular contraction.

## Neuron (Nerve Cell)

The cell body, dendrites, axon, and terminal branches of the axon are the four main components of all neurons. The nucleus is located in the cell body, which is densely packed with golgi apparatus and incoming projecting dendrites that create synapses with other nerve cells. A single axon (or sometimes more than one) extends from the cell body for a distance of 1 mm to 1 m. Basket cells, pyramidal cells, granule cells, and Purkinje cells are among the many types of neurons found in vertebrate nervous systems (the first three types being named after their structural shape).

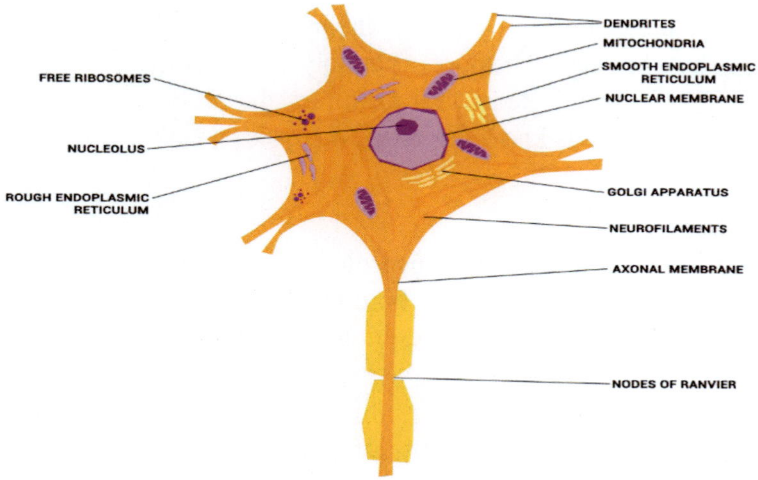


**Figure 41.** Labeled diagram of Neuron.

*Source: Image by Wikimedia Commons*

## Neuropathy

Behavioral or neurological illnesses, some of which are thought to have a genetic component.

## Neurosecretion

Neurohormones are produced in the neurological system by specific secretory cells.

## Neurosecretory Spheres

Electron-dense molecules with a diameter of about 0.1 m have been seen in electron micrographs of neuronal axons and are thought to be carried down the axon.

## Neurospora crassa

In genetics, the ascomycete fungus is commonly utilized. Neurospora, along with sordaria and yeast, is an essential organism in cell biology. They are eukaryotic microorganisms with a eukaryotic cellular structure, which allows for rapid growth and easy culture. Neurospora generates ascospores and develops as hyphae mycelium. Following meiotic division, the spores are placed in a linear row of eight, making it possible to analyze gene segregation by growing each spore separately and examining the expression of a specific gene locus. Because it examines the products and cross-over events of meiosis, this is known as tetrad analysis.

## Neurotoxin

Toxin that mostly affects the neurological system. Cobratoxin and abungarotoxin, which function by binding to acetylcholine receptors on motor endplates, are neurotoxins present in various snake venoms and other sources.

## Neurula

The neural axis has fully formed after emerging from the neural tube at this stage of vertebrate development. The gastrula stage comes before the neurula stage.

## Neutral Mutation

A mutation that gives the organism concerned neither a selective advantage nor a disadvantage. Most nucleotide substitutions result in neutral or near-neutral changes, according to M. Kimura's theory. Because there is no selective pressure against such mutations, they may spread in a population by accident. Many geneticists, on the other hand, disagree with kimura's interpretation and believe that neutral mutations are rare or non-existent, with the debate centered on how fine an advantage or disadvantage must be to be meaningful or detectable.

## Neutral Substitution

Amino acid substitution in a protein that has no effect on the molecule's function.

## Neutron Scattering

Technique for detecting and quantifying ordered structures within biological materials using neutrons released by a high-flux neutron source. Biological materials are typically tested in aqueous and deuterated solutions (i.e., solutions in which the hydrogen atoms have been replaced by deuterium), with the degree of deuterium concentration adjusted to match the scattering length density of one of the components. The nucleosome's complicated structure has been investigated in this fashion, with the DNA and histone components matched in separate runs.

## Neutrophil

White blood cell polymorph with cytoplasmic granules that stain with both basic and acidic dyes These cells account for more than half of all human blood leucocytes. They have an amoeboid form, are phagocytic, and are especially prevalent during bacterial infection.

## Niche

A plant or animal's ecological status within a community of organisms, both in terms of contribution to and dependent on the available environment.

## Nicking

A phosphodiester bond on one of a native DNA molecule's two strands is broken. Physical or chemical damage, as well as certain enzymes, can cause nicking.

## Nicking-Closing Enzyme (Topoisomerase 1)

Normal DNA replication is aided by this enzyme. The *E. coli* enzyme is a monomer (M.= 110 000) that employs negatively supercoiled DNA as a template. Its purpose is to relieve supercoiling stress that would otherwise accumulate in the DNA during replication unwinding. This is accomplished by quickly nicking and then religating one of the two DNA strands after it has spun around the unbroken strand's axis. As synthesis progresses, this mechanism advances along the DNA molecule at the replicating fork.

## Nick Translation

*In vitro* method for homogeneous radioactive labeling of DNA. DNA in solution is combined with trace amounts of DNAase I to produce nicks, deoxyribonucleotide substrates, one or more of which may be radioactive. DNA polymerase attaches to the nicks that are randomly created, excises the current nucleotides, and replaces them with the radioactive ones given. The polymerase precisely replicates the unnicked second strand, yielding highly radioactive DNA (107-109 counts per minute per microgram) with an unaltered base sequence at the end of the procedure. Nick translation is a common technique for preparing probe DNA for molecular hybridization.

## Nicotinic Acid (Niacin)

Water-soluble vitamin pyridine carboxylic acid (member of the b group of vitamins). Pellagra is caused by a lack of nicotinic acid. It's the building block for the nicotinamide ring, which is found in Nad+ and Nadp+.

## Ninhydrin Reaction

Amino acids and proteins separated by paper chromatography are commonly detected using biochemical tests. In the presence of proteins or amino acids, triketohydrinadene hydrate (ninhydrin) when heated with the test chromatograph turns blue.

## Nissl Body (Nissl Granule)

Granules of ribonucleoprotein. Found in the cell bodies and dentrities of neurons, but not in the axon hillock, and likely involved with protein production.

## Nitrification

Soil microorganisms convert nitrogen from complex organic molecules to nitrate. Nitrates are a readily available food source for higher plants.

## Nitrocellulose Filter

A paperlike mat of nitrocellulose fibers that is commonly employed in molecular biology. Single-stranded molecules flow through the filter, whereas double-stranded nucleic acids stick to it selectively. In southern blotting of gels, filters are utilized.

## Nitrogen Cycle

Nitrogen exchange between living organisms and the environment, as well as between organisms and the environment. Nitrogen concentration in microorganisms that are either free-living or in symbiotic relationships with plants converts atmospheric nitrogen to ammonium ions and eventually organic nitrogen. Electrical discharges in storms combine some air nitrogen with oxygen, and the resulting oxides dissolve in rain to generate nitrous and nitric acids, which interact with mineral salts in the soil to form nitrites and nitrates. Furthermore, nitrogen is fixed artificially (e.g., the Haber-Bosch method) to produce nitrogenous fertilizers. Microorganisms excrete and break down decaying organic debris, returning nitrogen to the soil in the form of ammonium (ammonification). Microorganisms turn this back to nitrites and nitrates.

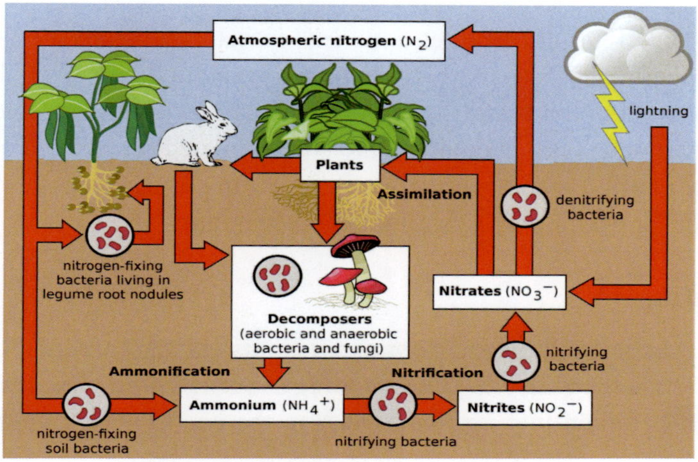


**Figure 42.** A diagram showing the Nitrogen Cycle.

*Source: Image by Wikimedia Commons*

## Nitrogen Fixation

Incorporation of atmospheric nitrogen into nitrates by symbiotic bacteria which live in the root nodules of leguminous plants. This nitrogen thus comes to be incorporated later into the tissues of the plant as a constituent of protein.

## Nitrogenous Base

Nucleosides, nucleotides, and nucleic acids are heterocyclic compounds that contain both carbon and nitrogen in a ring structure. The name is commonly associated with purine and pyrimidine bases, but it can also refer to other nitrogen-containing heterocyclic groups found in nucleotide coenzymes, such as flavin and nicotinamide.

## Nitrous Acid (Hn02)

By converting the purine and pyrimidine base amino groups to hydroxyl groups, this acid works as a mutagen on cells.

## NK Cell (Natural Killer Cell)

Some cultivated tumor cell lines and virus-infected cells are spontaneously and non-specifically targeted by a cytolytic vertebrate cell. The NK cell is thought to target a membrane glycoprotein with a modified carbohydrate; the ly5 differentiation antigen could be the NK cell receptor. NK cell activity is boosted by interferon and macrophage-stimulating drugs. Invertebrates may have NK cells, which play a role in natural immunity.

## Non-CodingDNA

DNA that is either in the nonsense strand (negative strand) of the DNA duplex or is not associated with any gene's coding sequence. Intron sequences may or may not be taken into account. DNA in the centromeric region is non-coding.

## Non-Darwinian

The process of evolution in which population changes are caused by circumstances other than natural selection.

## Non-Histone Proteins

Proteins that are not his tones and are located in the nuclei of eukaryotic cells. These proteins are acidic, unlike positively charged histones, and include the hmg protein as well as a wide range of polymerases and gene regulatory proteins required for transcription control.

## Non-Mendelian

Due to processes such as gene conversion, the ratios in the progeny of a cross do not follow Mendel's principles.

## Non-Polar

Chemical groups that lack the polar group required for aqueous interaction and so are not water-soluble.

## Non-Random Mating (Assortative Mating)

Because of some mechanism of selection or specialization at work, the chances of a male gamete fertilizing female gametes in the population are not random.

## Nonsense Codon (Stop Codon, Termination Codon)

Any of three codons that do not code for an amino acid and instead cause protein synthesis to stop. They are UAG, UAA, and UGA, with UAG representing amber, UAA representing ochre, and UGA representing opal.

## Non-Transcribed Spacer

A non-transcribed DNA segment that acts as a spacer between tandem copies of a transcribed gene sequence. It's worth noting that the Xenopus non-transcribed spacer between the copies of the major ribosomal genes is now known to be transcribed.

## Nopaline

When present in a crown gall, one of the opine compounds synthesized by plant cells stimulate the agrobacteria, causing the gall to grow rapidly.

## Noradrenaline (Norepinephrine)

The adrenal medulla releases an adrenaline-like neurohormone that promotes vasoconstriction and raises blood pressure. It's also found in synaptic vesicles in smooth muscle and glandular tissue in the sympathetic nervous system, and its release is significantly calcium-dependent both here and in the adrenal medulla.

## Northern Blotting

Procedure. Similar to southern blotting, however, the nucleic acids transferred are RNAs rather than DNAs. Capillary action transfers nucleic acids separated into subtractions by agarose gel electrophoresis to a nitrocellulose filter. They are then heated to adhere to the sheet and probed with radioactively labeled nucleic acids before being autoradiographed. It's worth noting that, although southern blotting was named after its originator,Southern, northern blotting was merely given a different name.

## Nucellus

An ovule's central tissue, comparable to the megasporangium. The megasporocyte is formed by a single nucleus cell, with the remainder cells serving as nutritive cells. In some species, the nucellus, also known as the periderm, serves as a food storage for the developing embryo.

## Nuclear Cage (Nuclear Lamina, Nuclear Matrix)

The fibrous layer just inside the nuclear envelope's inner membrane. This laminar structure is connected to all nuclear pore complexes as well as numerous chromosome DNA loops. It provides the nucleus with a skeletal structure and may also provide continuity between actively transcribing chromatin and the nuclear pores themselves.

## Nuclear Emulsion

The paths of ionizing molecules during radioactive decay are visualized using a photographic emulsion. Autoradiography employs such emulsions.

## Nuclear Envelope

The eukaryotic nucleus is surrounded by a two-membrane envelope. The envelope's two membranes are frequently separated by a significant distance, resulting in a specific perinuclear cisterna. Nuclear pores pass through the envelope, which is sometimes studded with ribosomes on its outer cytoplasmic surface.

## Nuclear Magnetic Resonance Spectroscopy (NMR Spectroscopy)

Technique used in chemistry and biochemistry to identify different compounds in complex mixtures and to quantify the amounts present by Analysing absorption spectra following changes in the resonance frequencies of the atomic nuclei.

## Nuclear Pore

One of hundreds or thousands of differentiated structures found in eukaryotic cells' nuclear envelopes. Pores are commonly found in clusters on the nuclear surface, and instead of being simple holes, each one has an annulus within it that fills or partially fills the hole. Although the size of pore annuli may control some of the selectivity of transport into and out of the nucleus, other effects require active transport and are tp-dependent. The nuclear pore complex is the name given to the complete structure, which has an inner diameter of 80 nm and an outer diameter of 100 nm. The annulus is made up of eight giant protein granules organized in an octagonal sequence. The pore has a small hole through it.

## Nuclear Transfer (Nuclear Transplantation)

Micromanipulation using fine glass needles allows nuclei to be exchanged between cells. Nuclei can be exchanged between big protozoan cells (e.g., *Amoeba proteus*) to give one cell two nuclei. A nucleus from a cell at one stage of the cell cycle is transplanted into an enucleated cell at a later stage. Nuclear transfer occurs in higher organisms when somatic nuclei from differentiated tissues are inserted into enucleated egg cells. A population of cloned individuals can be created by injecting many nuclei from one individual into a succession of egg cells.

## Nucleases

Nucleic acid degrading enzymes: DNase degrades DNA and RNAse degrades RNA. Exonucleases are enzymes that cleave nucleic acid polymers at their ends. Endonucleases and restriction endonucleases both cut within the polymer, however, restriction endonucleases only cut certain sequences.

## Nucleic Acid

Only two nucleotide polymers (DNA and RNA) occur naturally, while hybrid polymers between the two nucleic acids can develop momentarily or through manipulation.

## Nucleic Acid Hybridization

A hybrid duplex polynucleotide is formed by joining complementary strands of nucleic acid and forming hydrogen bonds between the appropriate pairs of bases. DNA and DNA, DNA and RNA, and RNA and RNA are all examples of hybridization. Despite the fact that such association is an important part of natural cellular architecture and metabolism, the word hybridization is usually reserved for circumstances in which the two complementary strands come from different origins and the relationship is regulated in vitro. Thus, probe DNA from one source is used to challenge the DNA run on a southern blot, which is taken from a separate source. In situ hybridization occurs when a probe is used to hybridize with DNA in fixed cell cultures.

## Nucleolus (Plasmosome)

The eukaryotic nucleus is divided into several parts. Ribosomes are formed in this area, which can be seen using optical or electron microscopy. Although frog oocytes can have over 1000 nucleoli, most somatic cells have two to seven nucleoli in their nucleus. The nucleolus is made up of granular (newly formed ribosomes) and fibrillar (DNA, including ribosomal cistrons) areas that are not separated by a membrane.

## Nucleoprotein

Protein and nucleic acid aggregation. Many of the essential components of eukaryotic cells, such as ribosomes and chromosomes, are formed by such aggregates. Because nucleic acids, including DNA, have a net negative charge. Its phosphoric acid residues are largely to blame. It is primarily connected with basic proteins that neutralize the charge, such as histones and protamine.

## Nucleosome (Nu Particle)

Eukaryotic chromatin's fundamental unit, made up of eight histone molecules and around 200 base pairs of DNA. Nucleosomes are strung together like beads on a string in chromatin, with the precise arrangement of the beads depending on the degree of chromatin compaction. Nucleosomes are assumed to be tightly packed in a solenoid pattern in compact chromatin, forming a Fiber roughly 30 nm in diameter. However, when the hl histone is decondensed, its conformation is altered, and roughly 60 base pairs of DNA form a linker between nucleosomes. Each core unit has two turns of DNA that are around 140 base pairs long. The core particle's histones are made up of two molecules of each of the H2a, H2b, H3, and H4 histones.

## Nucleus

Most eukaryotic cells have a spherical or lobular form. Nucleoplasm, nucleolus, and chromatin are contained within a nuclear envelope. At some periods of the cell cycle, chromatin can be separated into individual chromosomes. Most cells have a single nucleus; however, some non-cellular plasmodia may be momentarily binucleate. Some protozoan cells, such as paramecium, have two nuclei with different sizes and functions. Micronuclei and macronuclei are two types of nuclei. When mature, a few eukaryotic cells, such as mammalian erythrocytes, become anucleate. The nucleus is the center of transcription and RNA production in eukaryotic cells, as well as the main storage of genetic material. The RNA is solely translated into protein in the cytoplasm.

## Null Hypothesis

Statistical method based on the premise that there is no difference between the populations from which samples are collected, for example. The likelihood of finding a difference equal to or higher than the one seen is then calculated, and if it is low—less than 0.05--the null hypothesis is rejected, and the differences in the populations are evaluated as significant.

## Nurse Cell

A cell found in the ovary of insects that feeds the developing oocyte. The oocytes proliferate at the expense of the nurse cells because their cytoplasm flows into them. The cycles of DNA replication in nurse cells are endomitotic (i.e., No nuclear division occurs) (i.e., No nuclear division occurs).

## Nystatin

Fungicide used to treat fungal infections in the mouth, intestines, and urogenital tract. Nystatin can be taken orally or used topically. When taken orally, however, nystatin is not absorbed and hence cannot be utilized systemically. Nystatin can also be used to suppress fungal development in eukaryotic cell cultures; fungicides are normally unnecessary in bacterial cultures since the bacteria outgrow any fungal contamination.

# O

## Occam's Razor

The simplest appropriate hypothesis is favored in science; named after the scholar William of Occam in 1347. When numerous plausible explanations for a phenomenon are presented, the one that is preferred involves the fewest assumptions and is the most compatible with the data given. The parsimony principle is another name for it.

## Ochre Mutation ( Oc)

The codon UAA is the result of a nonsense mutation that alters the nucleotide sequence of a gene. Polypeptide biosynthesis is prematurely terminated as a result of this.

## Octopine

Plant growth ingredient (opine) produced by crown gall cells on plants and active in stimulating agrobacterium growth in crown gall cells.

## Odontoblast

Dentine synthesis is thought to be controlled by a cell found in the pulp cavity of a tooth. Long cytoplasmic projections that penetrate into the surrounding dentine distinguish odontoblasts.

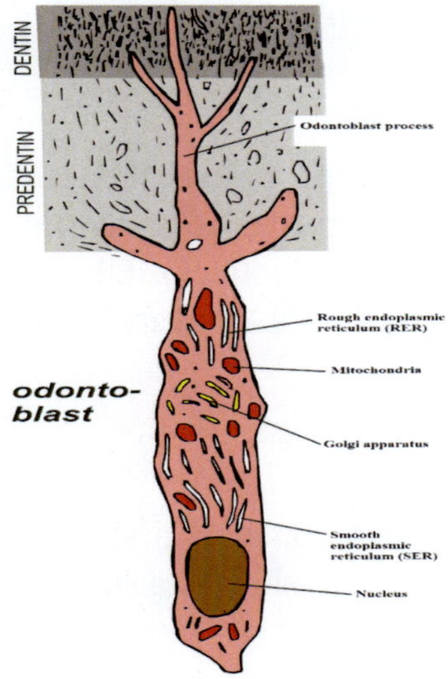


**Figure 43.** A diagram of Odontoblast.

*Source : Image by Wikimedia Commons*

## Oestrogens

Group of steroid hormones synthesized mainly by ovary and placenta, but also by testes and adrenal cortex, using acetate and cholesterol as precursors. Their main function in vertebrates is to stimulate the growth and maintenance of reproductive organs and some secondary sexual characteristics. Estrogen secretion from mature graafian follicles is regulated by pituitary gonadotrophins. Oestrogen hormone receptors have been identified on the surfaces of cells in responding tissues, and like other steroids, estrogens enter the target cells and stimulate specific genes into increased transcriptional activity.

## Ohno's Law

Because of suggested unique regulatory properties of the X chromosome, genes present on the X chromosome in one species are likely to be X-linked in other species; as suggested by S. Ohno.

## Oil Immersion Objective

With a drop of oil between the lens and the cover slip of the slide being studied, this light microscope objective lens can be used. Oils that closely match the refractive index of the glass are employed, slightly increasing the resolution of the light microscope.

## Okazaki Fragment

Prokaryotes (where they average several thousand nucleotides in length) and eukaryotes both have short pieces of newly generated DNA (where they average only a few hundred nucleotides). Okazaki fragments are formed during the normal replication of DNA and are quickly ligated *in vivo* to form whole chromosomes.

## Oligodendrocyte

The vertebrate nervous system contains glial cells. In the nervous system, these cells coil insulating myelin sheaths around axons and other neuronal processes.

## Oligosaccharide

Carbohydrate with a limited number of subunits. These molecules usually make up the carbohydrate moiety of glycoprotein.

## Oncogenes

Genes that have the potential to cause cancer. Although they are not native to retroviruses, they were first detected in them. Oncogenes are genes that are essential for normal growth and development and are found in both malignant and non-cancerous cells. Unlike normal cellular genes, cellular oncogenes, also known as proto-oncogenes, have a split structural organization with exons and introns, whereas viral oncogenes do not. Protein kinases or protein phosphorylating enzymes are the protein products of many of the oncogenes examined.

## Oncogenic Viruses

Viruses capable of transforming cells in tissue culture and inducing cancer in living animals. Oncogenic viruses include both RNA and DNA viruses; one category of RNA viruses, the retroviruses, has a large number of oncogenic members, but just a few herpesviruses, adenoviruses, and papovaviruses are oncogenic. The presence of an oncogene is certainly responsible for many oncogenic viruses' cancer-inducing capabilities, while the precise significance of such genes in cancer induction is unknown. Some, like the Epstein-Barr virus (eb), are only carcinogenic when other helper viruses are present. Although it is suspected that leukemias, breast and cervical carcinomas, lymphomas, and other human malignant diseases have a viral connection, this remains uncertain.

## Ontogeny

Individual organism's growth and development from fertilization to adult maturity.

## Oogenesis

The ovum is produced from the oocyte. The primary oocyte in the ovary experiences its first meiotic division to produce a secondary occyte with the major cytoplasm and a first polar body with minimal cytoplasm (during which prophase is halted and development and maturation occur before the division is finished). The ovum and a secondary polar body are produced by the secondary oocyte's second meiotic division. At ovulation, one or both oocyte stages are released and fertilized without the ovum being produced in some species.

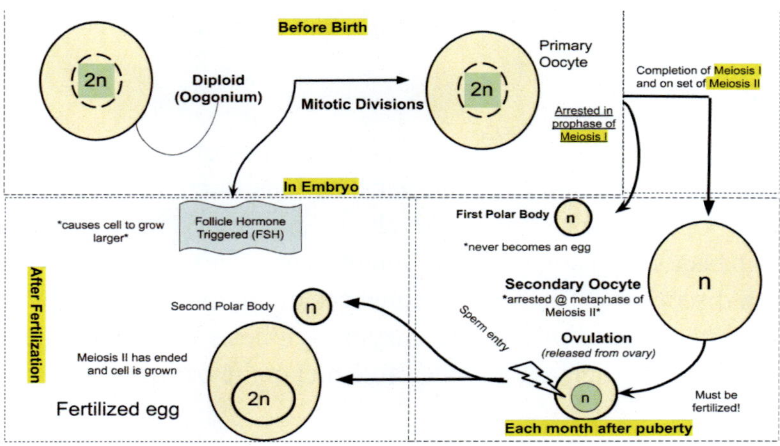


**Figure 44.** A labeled diagram of Oogenesis.

*Source: Image by Wikimedia Commons*

## Oogonium

A cell that divides to create oocytes in the female ovary. It is the basic female sex organ of algae and fungi, with one too many oospheres.

## Oosphere

In plants, a naked female gamete with some nutritional stores. It is frequently produced in an oogonium, but not always.

## Opal Mutation

Nonsense mutation that causes the codon UGA to be added to a gene's nucleotide sequence, signaling the end of polypeptide production.

## Open Reading Frame (ORF)

A region of a DNA sequence that can be interpreted as codons specifying amino acids (i.e., It is not interrupted by nonsense codons as stop signals). ORF sequences are regularly discovered in gene bank sequencing and can only be attributed to a specific known gene in rare occasions.

## Operator

A segment of DNA capable of interacting with a specific repressor molecule and, as a result, regulating the activity of other genes downstream.

## Operon

Operons are a type of DNA sequence that consists of two or more coding sequences and a regulatory region that functions to coordinate the coding sequences. Operons can be negatively or positively regulated and are frequent in bacterial genomes. In the genomes of eukaryotes, no real operons have been discovered. The coding sequences of operons, which account for the coordinated pattern of induced enzyme synthesis in bacteria, generally specify enzymes participating in the same metabolic pathway.

## Opine

Crown gall cells produce this compound, which the agrobacteria in the gall use as a growth medium. Opines include nopaline and octopine, for example.

## Opine Synthetase (Opine Synthase)

The enzyme system responsible for the synthesis of opines, which come in a variety of forms. The T-DNA region of the plasmid contains the genetic material for opine synthetase. These plasmids provide agrobacterium strains that cause crown gall disease virulence. The T-DNA is transferred to plant cells and stably integrated into the nuclear genome, where the opine synthetase gene is transcribed and translated.

## Opsonins

A class of molecules that cause phagocytosis in cells. Because macrophages contain Fc receptors on their surfaces, they can bind antigen-antibody complexes, and antibodies like IgG and IgM work as opsonins, especially when complement is present.

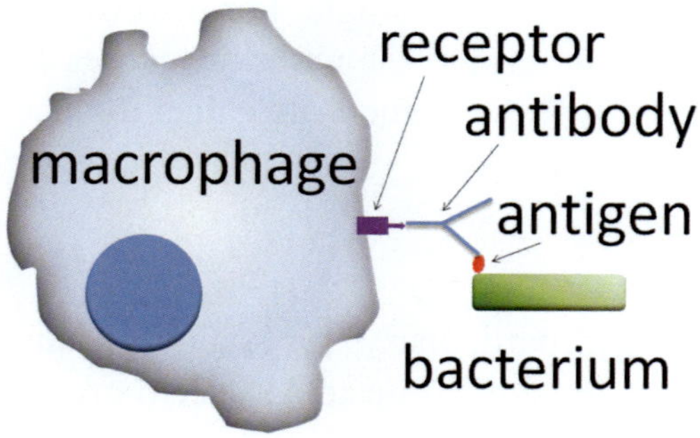


**Figure 45.** A labeled diagram of Opsonins .

*Source: Image by en-Academic.com*

## Optical Activity

When polarized light passes through a molecule's solution, it has the ability to spin the plane of the light. Monosaccharides, for example, can take one of two isomeric forms. They occur in two configurations known as d-form (dextrorotatory) and l-form (laevorotatory) as a result of their asymmetric carbon atoms, bending light to the right or left, respectively.

## Optical Density (Od)

The degree of light absorption by a solution as a function of the solute content. The relative optical density of light-absorbing solutions offers information about the concentration and the character of the solution. Spectrophotometers are devices that measure optical density using visible light, infrared light, or ultraviolet light. Scanning spectrophotometers produce an absorption spectrum over a range of wavelengths, which can be used to detect a solution's exact composition and purity. Proteins absorb light at a wavelength of 280 nm, whereas nucleic acids absorb light at 260 nm. The protein content of an impure DNA solution can therefore be determined. Changes in optical density can be utilized to provide precise information about DNA complexity, notably in the usage of reduction in optical density after melting, as is used to produce cot plots, because the optical density of a DNA solution increases when the DNA is melted (hyperchromic effect).

## Optical Isomers

When a beam of plane-polarized light passes through a molecule's solution, the molecule's isomers cause it to rotate. For distinct isomers of the same molecule, the asymmetric form of the molecule causes rotation to the right (dextro-) or left (laevo-). These isomers are called d-glucose or l-glucose in the case of glucose, for example.

## Orcein

Dye used in cytology, particularly for staining squash chromosomal preparations when dissolved in acetic acid as aceto-orcein.

## Ordered Tetrad

In the row of eight ascospores of a fungal ascus, the products of meiotic division are observed in a linear sequence. Growing each spore independently and scoring it for expression of one allele at a specific locus can be used to study crossing-over events.

## Organ Culture

In contrast to tissue culture, which includes the growth of cells without a multicellular morphological structure, organs or portions of organs grow in culture. The culture of small embryonic gland primordia (e.g., salivary gland) or skeletal primordia is best exemplified for complete organs. Organ culture is usually only kept for a short time. Because of the restrictions of food diffusion from the culture medium, prolonged growth leads to a point where tissue necrosis develops centrally, limiting the technique's utility.

## Organic

(1) As applied to living or dead organisms, as well as time-derived compounds. (2) A term for chemical molecules built on carbon rings or chains, even if they have no direct biological relevance. Oil and its derivatives are hence organic compounds.

## Organogenesis

Organ development during embryonic life. It includes both individual cell differentiation and the structural form of the organ as it develops (morphogenesis).

## Origin Of Life

Any theory that suggests life evolved from non-living processes, whether on this planet or on another analogous body in the universe. Rather of jiving cells constantly originating from non-living systems, most explanations think that life systems evolved at a specific moment in the evolution and history of the earth. Evidence from morphological remains of microbes and extrapolation from current biological forms suggest that the genesis of life occurred approximately a thousand million (or billion) years ago. The atmosphere at the time was thought to contain ammonia, methane, water vapor, and hydrogen, but no oxygen. Lightning discharges or ultraviolet radiation could have produced amino acids and even peptides on a substrate containing simple molecules. Photosynthesis (which produces oxygen) almost definitely occurred early, whereas aerobic respiration would emerge only when oxygen was abundantly available. Early cellular forms are thought to have died from exposure to oxygen. It's still a mystery how early biological systems formed cell-replicating cellular units. Even primitive prokaryotes are incredibly complicated, and the earliest progenitors of living cells would have looked a lot like archaebacteria today.

## Ornithine Cycle (Urea Cycle)

The nitrogenous products of protein metabolism are converted and detoxified in a cyclic series of metabolic processes. The metabolic route produces ammonia and urea, with the former being used to convert ornithine to citrulline and the latter being mostly eliminated. Some human genetic mutations cause blockages in the cycle, resulting in citrullinuria and ornithine transcarbamylase insufficiency, for example.

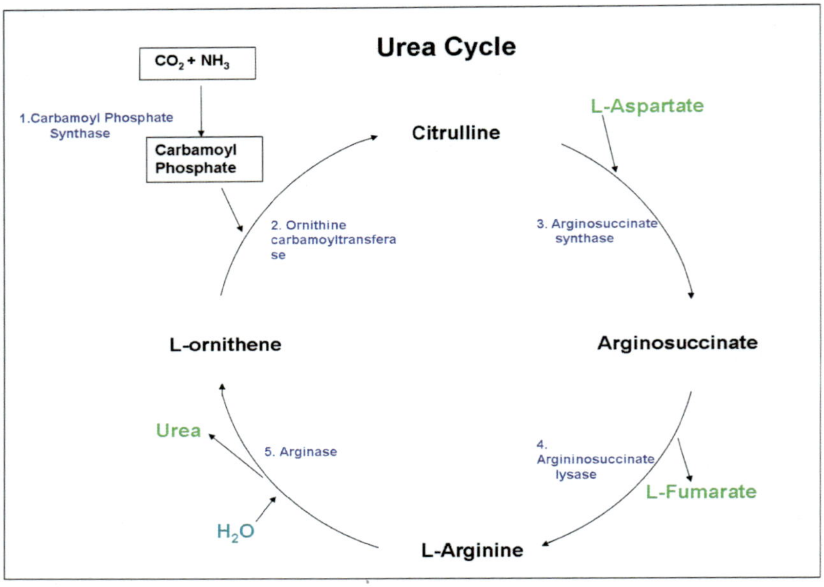


**Figure 46.** A diagram of the Ornithine cycle (Urea cycle).

*Source: Image by Wikipedia*

## Orthogenesis

Evolution followed certain paths, specified by inherent properties of the genotype, rather than genetic drift and natural selection, according to a now-discredited theory.

## Osmosis

The spontaneous movement of a solvent from an area in a solution where its activity is high. To a place where it is less active. In practice, osmosis occurs only when a semipermeable membrane separates a solution from a pure solvent or a solution with the same solvent but a different solute concentration. Water is the solvent in biological systems, and its flow tends to balance solute content on both sides of a cell membrane. However. Cell membranes are not truly semipermeable because they allow some solutes to pass through, hence they are referred to as selectively permeable.

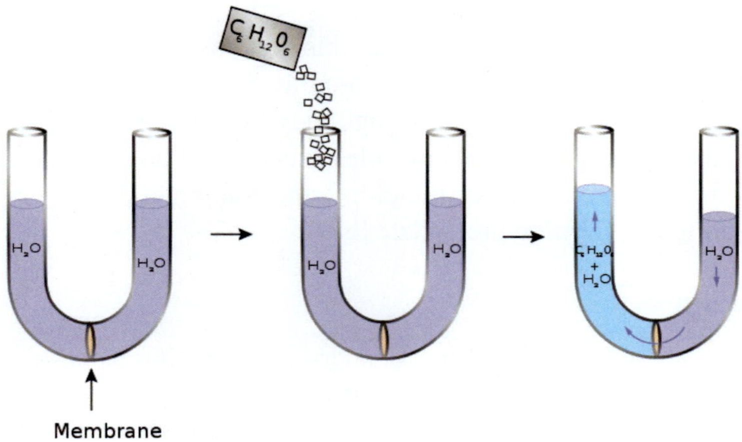


**Figure 47.** A diagram of Osmosis.

*Source: Image by Wikimedia Commons*

## Ossification

Bone development. It is a procedure that involves the replacement of pre-existing connective tissue with cartilage (endochondral ossification) or a simple connective tissue membrane (intramembranous ossification). Osteoblasts infiltrate connective tissue, secreting an extracellular matrix of bone before being imprisoned as osteocytes in the matrix. Some resorption of existing bone may occur when existing bone is reformed in development or repair, which is performed by osteoclast activity.

## Osteoblast

Bone-forming cell that deposits extracellular matrix and gives rise to osteocytes on the surface of bone matrix.

## Osteoclast

Extracellular enzymic activity is responsible for the resorption of bone matrix by multinucleate giant cells. Osteoclasts are distinguished from osteoblasts and osteocytes by the fact that they are generated by the fusing of mononucleate progenitor cells called monocytes.

## Oudin Technique

Similar to the ouchterlony technique, this test involves the synthesis of precipitation zones as a result of antigen-antibody interaction. A gel column with an even dispersion of a specific antibody molecule is created. After an aqueous suspension of antigen molecules is deposited on the upper layer of the column. Antigen molecules travel into the column and bind selectively with the antibody, forming precipitate lines.

## Ovalbumin

One of the most important proteins found in egg white. It is a phosphoprotein with 386 amino acids that is the main nutritive protein in avian egg white.

## Overdominance (Monohybrid Heterosis)

Heterozygotes may show a larger phenotypic extreme of gene expression than homozygotes with the same allele.

## Overlapping Code

In general, the genetic code is non-overlapping, with each triplet of bases coding for a different amino acid. Furthermore, adjacent triplets do not share common bases. Some overlapping genes do show how the same sequence can be used with a different reading frame. This results in triplet overlap.

## Overlapping Genes

Genes with some overlap in their nucleotide coding sequences. Apart from some regulatory genes in bacteria that overlap. The only unambiguous example known is in the genes of certain phages, such as the operator and promoter sequences of the tryptophan operon of *E. coli*. One gene, for example, is wholly within the coding sequence of another in the phage xl74 genome, but is translated using a different reading frame.

## Overwinding

The DNA double helix is already twisted in a righthanded spiral in this situation. Is subjected to extra torsional stress in the same direction, causing the double helix to positively supercoil on itself.

## Oviparous

Describes organisms that lay eggs, develop embryos, and hatch them outside of the female's body.

## Ovotestis

In some hermaphrodite species, the reproductive organ consists of both an ovary and a testis (a simultaneous hermaphrodite). Alternatively, at regular intervals (a sequential hermaphrodite). There are species in the phylum Mollusca that display one or the other arrangement.

## Ovovivipary

The formation of distinct eggs is required for the reproduction of some animal species. Although no direct contact between the egg and the female is formed, the embryos develop within the egg inside the female's body. Ovoviviparous animals include fish and reptiles.

## Ovulation

In female vertebrates, an ovum or egg is released from a mature follicle of the ovary. The hypothalamic releasing factors move down the pituitary portal veins to the pars distalis of the pituitary, where they stimulate pituitary release of luteinizing hormone and follicle-stimulating hormone, which together promote ovulation in animals.

## Ovule

Structure in seed plants that houses the egg cell and develops into the seed after the egg cell is fertilized.

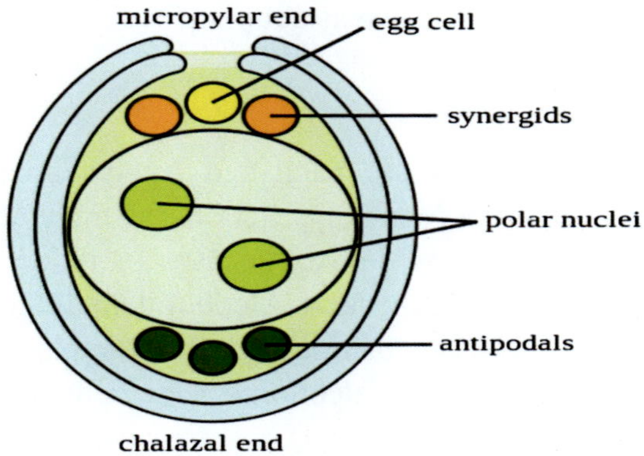


**Figure 48.** A labeled diagram of Ovule.

*Source: Image by Wikipedia*

## Oxidases

Enzymes that catalyze oxidation processes with oxygen as the electron acceptor (oxidant), water or hydrogen peroxide (h20 2) as the product, and no oxygen atoms integrated into the product. Amino acid oxidases, for example, are flavoproteins that catalyze the oxygen-dependent oxidation of amino acids to keto acids while also producing hydrogen peroxide. The cytochrome oxidase, which reduces oxygen to water using electrons from an electron transport chain and protons from the surrounding medium, is perhaps the most important example of an oxidase.

## Oxidation

The addition of oxygen, or the joss of hydrogen, or the reduction of the number of electrons in an element's or compound's molecular structure.

## Oxidation/Reduction Reaction

A chemical reaction in which electrons are transferred from one molecule to another. The atom or molecule that loses an electron is called the reductant, and it is oxidized in the reaction, whereas the atom or molecule that gains an electron is called the oxidant, and it is reduced.

## Oxidative Phosphorylation

Phosphorylation of ADP to ATP in aerobic respiration (as in the electron transport chain of mitochondria). Uncoupling drugs like oligomycin and dinitrophenol suppress phosphorylation, which is controlled by proteins in the mitochondrial membrane known as coupling factors.

## Oxidoreductases

Oxidation/reduction processes are catalyzed by enzymes. Because all oxidizing processes require a reducing reaction, the enzymes that catalyze them (commonly referred to as dehydrogenases or reductases) are more correctly referred to as oxidoreductases. Alcohol dehydrogenase, aldehyde reductase, or alcohol: NAD+ oxidoreductases are examples of enzymes that catalyze the reversible oxidation of alcohol to acetaldehyde utilizing NAD+ as a cofactor.

## Oxygenases

Enzymes that catalyze (typically irreversible) oxidation processes in which oxygen is integrated into the product from molecular oxygen. In the metabolism of stable reduced molecules like alkanes, oxygenase-catalyzed reactions are frequently the first oxidative step. Monooxygenases catalyze the incorporation of a single atom of oxygen, with the second atom being reduced to water by an electron donor (reductant), commonly NADPH, in the same process. Between the electron donor and the substrate, monooxygenase processes frequently involve a number of electron transfer components, such as iron-sulfur proteins, flavoproteins, or cytochromes. The mechanism for detoxification of foreign substances in mammalian livers involves a cytochrome called cytochrome P450. Because monooxygenase processes typically result in a molecule with an additional hydroxyl group, they are generally referred to as hydroxylation reactions. Dioxygenases differ from monooxygenases in that they integrate both atoms of oxygen from molecular oxygen into the product and do not require a reductant.

## Oxygen Quotient (Q02)

Figure depicting the rate at which a tissue or organism consumes oxygen, represented in microliters of oxygen per milligrams dry weight per hour.

## Oxytocin

The mammalian pituitary gland secretes a polypeptide hormone. It induces smooth muscle contractions, particularly in the mammary glands' alveoli during lactation and the uterine wall during parturition (birth).

# P

## Pachytene (Pachynema)

Following leptotene, the chromatid threads of the chromosomes split, save at the centromere, resulting in each bivalent chromosome being converted into a tetrad.

## Packing Ratio

The degree of compaction of the DNA double helix into the chromatin fibers that encapsulate it is expressed as a ratio. Each chromosome of the human complement of 23 in the haploid genome comprises 5 mm of uncoiled DNA, which is reduced to around 1 mm of 30-nm diameter chromatin Fiber, which is further compressed by the higher-order structure of chromatin into a chromosome of the average length of only 6 11m.

## Paedogenesis

A type of neoteny in which an organism reaches sexual maturity while remaining morphologically a larva. The axolotl and the fly Miastor are examples of paedogenetic organisms, in which the larva (grub) generates the following generation without maturing further.

## Paedomorphosis

Adult forms of a species may mimic the premature stages of ancestor species, a phenomenon in evolution. In some circumstances, neoteny is thought to be at least partially responsible, and the human has been dubbed a neotenous ape.

## Palindromic Sequence (Inverted Repeat Sequence).

The code reads the same from both ends on complementary strands of a nucleic acid sequence. Such sequences contain a central axis of symmetry and can be folded in different ways to produce distinct tertiary structures with sets of base-paired loops that regulatory molecules can recognize as target sites. Restriction endonucleases attack palindromic sequences as well, albeit most do not cut directly across the axis of symmetry. In DNA reassociation experiments, such as those used to develop cot plots, such DNA hybridizes by folding back on itself, resulting in fold back DNA. In ideal conditions, hybridization occurs almost instantly.

## Pangenesis

Small molecules from various regions of the body, known as pangenes, were accumulated in gametes and transferred to children, according to an outdated theory of development. The two parental lines would have inherited a pattern of blending inheritance.

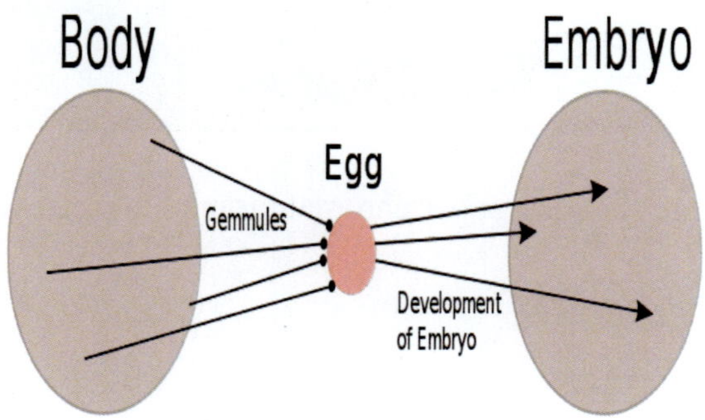


**Figure 49.** A labeled diagram of Darwin's Pangenesis.

*Source: Image by Wikimedia Commons*

## Pantothenic Acid

B complex vitamin found in abundance in yeast, liver, peas, and other leguminous seeds. It is a water-soluble molecule that is a precursor to coenzyme A and is necessary for the survival of many vertebrates, including humans.

## Paper Chromatography

Separating chemicals in a mixture on a filter paper strip is a method for identifying them. Near one end of the paper, the mixture is deposited as a spot. The strip is then dipped into a solvent reservoir at the other end. The molecules in the mixture migrate as the solvent flows along the paper via capillary action and past the spot. Each substance's rate of migration is influenced by its solubility in the solvent. As they travel down the paper, compounds with various solubilities are separated.

**Figure 50.** A picture showing paper chromatography in process .

*Source: Image by Wikipedia Commons*

## Papilloma

Benign neoplasm of the skin, many of which are caused by papillomaviruses infecting the skin cells. The common wart is a papilloma, and its non-invasive nature is revealed by the fact that the wart's cells do not grow into the subcutaneous layers, instead of piling up on the surface.

## Papillomaviruses

PAPOVAVIRUSES that cause papillomas in a variety of mammalian species. These viruses have double-stranded DNA genomes and are regular icosahedral units with a diameter of about 55 nm.

## Parabiotic Twins

As a result of surgery, two animals are connected together through functional connections. Parabiotic twins are frequently utilized to see if cells or molecules missing from one twin may be replaced with those from the other.

## Paracentrotus lividus

Common sea urchin: frequently utilized in developmental biology and molecular genetics research. It grows in abundance throughout North America and is purplish in color, it rarely exceeds 5 cm in diameter, Sea urchin sperm, which contains more than 50% pure DNA, is abundant, and it has thus become a significant source of DNA in research on this molecule.

**Figure 51.** A picture of *Paracentrotus lividus.*

*Source: Image by Wikipedia Commons*

## Paracrystalline Aggregate

Individuals with sickle cell anaemia has a regular array of molecules, such as Stored RIBOSOMES, VIRUS clusters, or hemoglobin molecules within sickled cells.

## Paramecin

Paramecium killer strain cells secrete a factor with the ability to kill non-killer, sensitive Paramecium cells. Kappa molecules, symbiotic bacteria of the species Caenobacter taenospiralis, synthesize the factor. Some data suggests that the paramecin factor is a DNA bacteriophage.

## Paramecium

Ciliated Protozoan genus that is prevalent in freshwater and has been extensively explored in terms of nuclear/cytoplasmic relationships. The most studied species is *P. aurelia*, but *P. caudatum* and *P. bursaria* are also popular. Paramecium species have two nuclei per cell: a big polyploid vegetative nucleus called the MACRONUCLEUS and a small diploid non-transcribed MICRONUCLEUS that is only active during conjugation. The pattern of cilia in rows of kineties on the surface cortex of these cells is also notable, particularly the fact that artificial changes in the kinety pattern, induced by microsurgery, are passed down indefinitely to progeny, indicating that the pattern is not under the precise control of nuclear genes.

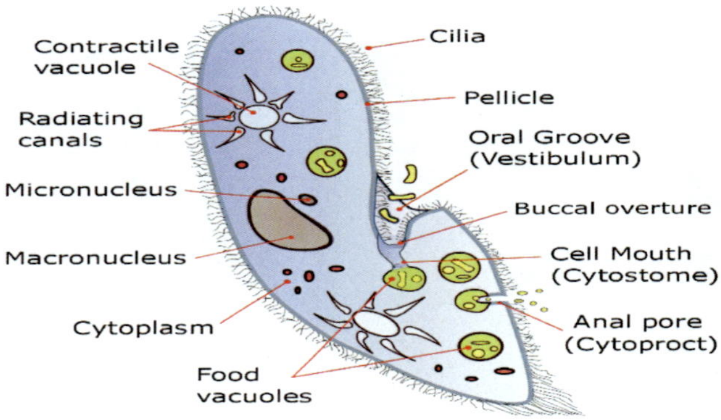


**Figure 52.** A labeled diagram of Paramecium.

*Source: Image by Wikimedia Commons*

## Parapatric

A population or species that has a common area of overlap in its geographical distribution, where hybridization occurs frequently.

## Parasexual Cycle

There is no meiosis in this reproductive cycle. Nuclear fusion may occur between two distinct nuclei in fungi, with subsequent recombination and generation of new nuclei with genetic assortments not seen in the original nuclei, because cells are often natural heterokaryons.

## Parasitism

The state of a relationship between two organisms in which one, the parasite, benefits from the other, the host, while the latter organism suffers as a result of the interaction.

## Parathyroid Hormone (Parathormone, PTH)

The parathyroid glands, which are located near the thyroid gland, secrete this hormone. It's a 75-amino-acid polypeptide hormone that raises blood calcium levels while also adjusting the calcium/phosphorus balance (the inactive precursor has nine more amino acids). When blood calcium levels drop, calcium is reabsorbed from bone and calcium salts are recovered by renal activity and gut absorption.

## Parenchyma

Most basic sort of differentiated cell found in plant tissue. The cells are usually thin-walled, vacuolated, and contain a protoplast, and their form is generally isodiametric. The ground tissue is made up of parenchyma, which can be found in the pith and cortex of stems and roots, as well as the mesophyll of the leaf. It's also found in the XYLEM and PHLOEM, both main and secondary.

## Parity

The fact of having given birth to a young child. A parous female has given birth to live children, and she is said to be of parity 3 if she has given birth to three groups of children, whether one or more each group.

## Parthenogenesis

Without fertilization, the development of an organism from an egg. It can be caused by physical or chemical artificial fertilization (activation) of the egg, the growth of unfertilized eggs into haploid individuals (as in some bee castes), or gynogenetic activation of the egg by sperm. Parthenogenesis is prevalent in plants and quite common in some animal species. Thus, it is frequent among rotifers (in certain species, males are unknown), and it explains the enormous increase in aphid numbers in the early summer.

## Passage Cell

Thin-walled cell in the endodermis that does not thicken its secondary wall but has a casparian strip. Passage cells are found opposite protoxylem poles and are thought to allow water and solutes to move from the cortex to the xylem. This name is also applied to small, thin-walled cells seen in the exodermis of epiphytic orchid aerial roots.

## Passive Immunity

The immunity provided through the acquisition of antibodies or sensitized lymphocytes from other immune animals rather than by the particular vertebrate's own immune system. In primates, the transfer of maternal IgG across the placenta imparts passive immunity to the fetus; IgA and/or IgG are passed from the mother to the infant in colostrum and milk in many mammalian species, and maternal immunoglobulins are found in bird eggs.

## Pasteurization

Pasteur developed a technique for eliminating unwanted germs from wine that is now commonly used for milk. The liquid is heated to 62°C for 30 minutes (or a fraction of a minute in 'flash' pasteurization at a higher temperature). Because many milk-borne infections do not develop spores and are reliably eliminated, the procedure is effective. Although the total bacterial count has been lowered by 99 percent, sterilization has not been achieved.

## Patch Clamping

A method for examining the action of ion channel proteins in the plasma membrane. A small region of membrane containing only one or two ion channels is permitted to bleb into the tube using a fine glass needle attached to the cell surface. Only the ion channels trapped inside the pipette allow current to enter, and transmission of low voltages with closed and open channels can be recorded using very sensitive electronic equipment.

## Patroclinous

A term used to describe a child who looks more like the father than the mother. The sex-linked genes are patroclinous in male offspring of female *Drosophila* with an attached X chromosome.

## Pattern Formation

Developing recognizable patterns of tissue organization in the embryo that may be serially repeated along the anterior/posterior axis (e.g., in a segmented animal), have bilateral symmetry (e.g., in left and right appendages), or simply be faithfully reproduced in successive generations of an organism (e.g., bristle patterns in insect cuticle, leaf venation patterns); often used as a complement to MORPHOGENESIS. One of the key challenges in developmental biology is the molecular/cellular foundation of pattern specification mechanisms.

## Pbr322

Plasmid from *E. coli* that is widely employed as a cloning vector to propagate other gene sequences. It can be harvested in big quantities since it grows in the cell under relaxed control. It has tetracycline resistance and ampicillin resistance genes.

## Pectic Substances

Many plants' cell walls have a mixed collection of acid polysaccharides. Long-chain molecules, some of which compose the middle lamella of plant tissue; others, particularly the pectin subgroup, are found in ripening fruit. They're commonly utilized in jam-making since they produce gels at low temperatures.

## Pedigree

Ancestral history, with emphasis on specific genetic features. Females are represented by circles and males by squares in pedigree diagrams, and individuals expressing a trait are represented by solid figures in a diagram for that trait.

## Penetrance

Proportion of individuals with a specific genotype who express a trait in their phenotype under certain conditions. Complete penetrance refers to when all individuals bearing a dominant gene manifest the mutant phenotype, whereas incomplete penetrance refers to when less than 100% of such individuals do.

## Penicillins

Antibiotics derived from the *Penicillium notatum* fungus and other Penicillium species They work by interfering with the manufacture of cell walls in gram-negative bacteria, causing many of them to become naked protoplasts and lyse as a result.

## Penicillin Screening

A technique for boosting the number of mutant cells in a bacterial population; principally used to select auxotrophic mutants. Following mutagen treatment, the bacteria are placed in a liquid minimal medium that supports the development of the parental wild type but not the auxotrophic mutant. Penicillin is added to the medium as well. In their endeavor to develop, wild-type cells are selectively killed by penicillin, whereas auxotrophs remain dormant and thereby escape the antibiotic's impact.

## Peptide Bond (Amide Bond)

The a-carboxyl group of one amino acid is joined to the a-amino group of another amino acid. The bond is formed when one of the two amino acids loses a water molecule, and the link can be dissolved by hydrolysis. The equilibrium favors hydrolysis; the opposite reaction, like protein synthesis, necessitates an input of energy and knowledge about the amino acid sequence. The carbonyl group in the peptide is generally linked to the amino group, and the bond is stiff and planar because it has a partial double bond character (owing to resonance).

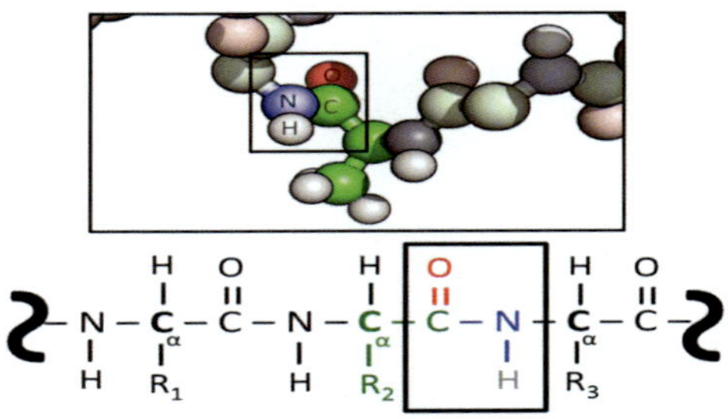


**Figure 53.** A diagram of the Peptide bond.

*Source: Image by Wikipedia Commons*

## Peptidyl Transferase

During the TRANSLATION Stage of protein synthesis, this enzyme catalyzes the movement of amino acid-charged tRNA molecules (aminoacyl-tRNA) on the ribosome. Prior to discharge, the charged tRNA molecules engage sequentially with the A and P sites on the ribosome, and the transferase assists the migration of the tRNAs from one site to the next.

## Pericycle (Pericambium)

A single layer of parenchyma cells that lies just within the endodermis and on the vascular tissues' periphery. The cells can divide and produce lateral root primordia and vascular cambium. The pericycle also gives rise to the phellogen, the meristem that generates cork.

## Periderm

Woody plant stems and roots, both angiosperms and gymnosperms, have an outer secondary tissue layer. It has a similar protective function to the epidermis and replaces it during secondary thickening. An outer phellem or cork layer is interrupted at intervals by lenticels; an inner PHELLODERM; and a central PHELLOGEN, or cork cambium. By cell division, the phellogen produces new phelloderm or phellem.

## Periosteum

Connective tissue layer that surrounds bone and serves as a point of attachment for muscles and tendons. The inner layers of the penosteum are loose and vascular, and it is here that osteoblasts develop before migrating into the bone proper. White elastic fibers make up the majority of the outer layer.

## Permeability

The ease with which molecules can permeate a membrane or porous surface. Semipermeable biological membranes allow some tiny molecules to pass through while excluding or partially excluding many larger molecules. diffusion causes the flow of numerous molecules through permeable barriers.

## pH

The acidity or concentration of protons (H+) in a solution. It is equal to the negative log10 of the hydrogen ion (proton) concentration in terms of numbers (or log10 of the reciprocal of the hydrogen ion concentration). As a result, the pH scale is a logarithmic scale. At pH 7.0, the quantities of hydrogen and hydroxyl ions are equal; solutions below pH 7 are acidic, and those above pH 7 are alkaline or basic.

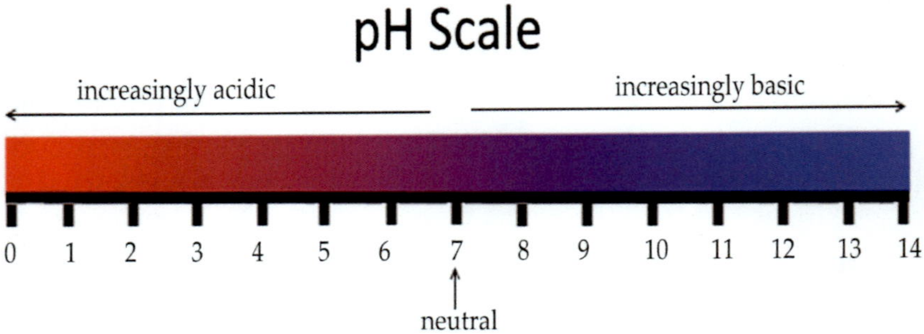


**Figure 54.** A picture of Ph scale .

*Source: Image by Wikimedia Commons*

## Phage Induction

In response to an external stimulation, a temperate phage transitions from the prophage stage to lytic growth. Induction causes phage genes to be transcripted, the prophage to be removed from the host chromosome, and phage genomes and capsid proteins to be synthesised.

## Phagocytosis

It's a type of ENDOCYTOSIS that involves the expansion of pseudopodia or filopodia to entrap extracellular molecules within membrane-bound vesicles; it literally means "cell eating." The particle almost completely fills the vesicle, which is huge and known as a phagosome. The particle is broken down after the phagosome fuses with a main lysosome to generate a phagolysosome. Phagocytosis can affect any mammal, although it is most commonly associated with specialized cells known as PHAGOCYTES.

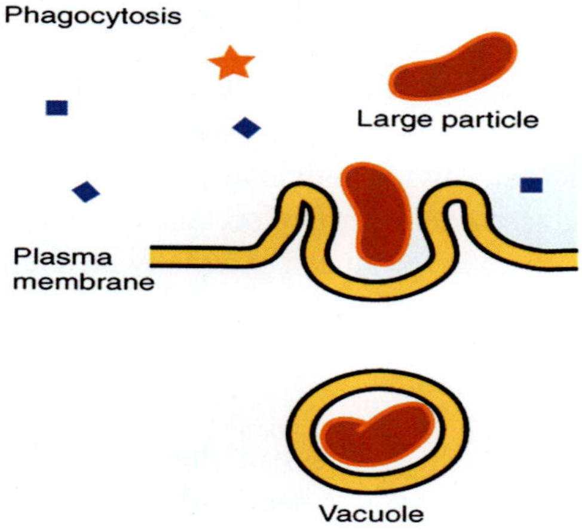


**Figure 55.** A labeled diagram of Phagocytosis.

*Source: Image by Wikimedia Commons*

## Phagosome

A cytoplasmic particle formed by the budding off of a portion of the plasma membrane and encapsulating recently phagocytosed material. They can only be found in phagocytic cells, where they will fuse with lysosomes to form phagolysosomes, or digesting vacuoles.

## Phellem (Cork)

It is formed from the phellogen and comprises of more or less isodiametric cells organized in radial rows. It replaces the epidermis in plants that undergo substantial secondary growth; it acts as a compact protective covering. The cells are empty of cellular contents and have thick cell walls impregnated with suberin when they reach maturity.

## Phelloderm (Secondary Cortex)

Internally located secondary plant tissue generated from the phellogen. The cells keep their live contents and are distinguished from cortical cells solely by their radial alignment with the phellogen and phellem. The PERIDERM is comprised of the phelloderm, phellogen, and phellem.

## Phellogen (Cork Cambium)

In stems that undergo considerable secondary thickening, the lateral meristem gives rise to the periderm, a protective layer that replaces the epidermis. phellogen cells divide periclinally to produce radially seriate cell files. internally produced phelloderm or secondary cortex differentiates into phellem or cork, while those produced externally differentiate into phellem or cork.

## Phenocopy

Environmental factors such as stress or diet cause the production of a specific phenotype. in a form that is nearly identical to or indistinguishable from that which is ordinarily produced by the expression of a specific gene or gene mutation

## Phenotypic Lag

Delay in expressing a newly acquired personality trait. This phrase usually refers to a recent mutation. Such a lag could be caused by a mitochondrial mutation, in which the presence of numerous mitochondria at first dilutes any evidence of phenotypic expression, or it could be caused simply by the cytoplasm carrying a large or long-lived store of the immediate gene product, and the effect of the mutant gene becomes apparent only when the store is depleted. Following the transplantation of nuclei across species in the unicellular alga Acetabu/aria, there is a phenotypic lag of many weeks.

## Phenotypic Mixing

A phenomenon caused by a cell being infected by viruses of many strains. This could result in the packaging of additional genetic material from one viral strain within a protein capsid characteristic of the other, resulting in the generation of complete virions with mixed phenotypes.

## Phenotypic Mutant

A mutant that has a distinct phenotype and may be classified according to it. the majority of phenotypic mutants are the result of analyzing features defined by additive factors, such as the drosophila apricot-colored eye, rather than the red eye, which is caused by variations in the pigment gene dosage. only allozyme analysis in electrophoresis can detect other phenotypic mutations.

## Pheromone

A chemical that causes another individual of the same species to behave differently. Pheromones include trail substances used by ants, territorial markers used by wolves, aggregation-promoting substances seen in bee swarming behavior, and sex attractants like the compound released by female emperor moths, detectable by males up to a mile away and forming a concentration gradient that the male uses to detect the female.

## Philadelphia Chromosome

Patients with chronic myelogenous leukemia have an aberrant chromosome that is visible in bone marrow metaphase. It is caused by a translocation, in which half of chromosome 22's long arm is transferred to the long arm of another chromosome, most commonly chromosome 9. because the breakpoint on chromosome 22 is close to an oncogene region, the resultant cancer might be linked to changes in the gene's expression regulation.

## Phocomelia

A single bone connects the hands or feet to the trunk in this condition. It's a rare human autosomal recessive trait, although it's most well-known in infants born to moms who took the tranquilizer thalidomide.

## Phosphodiester Bond

A nucleic acid polymer is formed by a linkage bond between nucleotides. It entails linking the phosphate group of one sugar's 5' position to the hydroxyl group of the sugar of an adjacent nucleotide's 3' position.

## Photosynthesis

The conversion of light energy into chemical energy, which is then used to manufacture organic molecules via carbon dioxide reduction. Except for chemosynthetic organisms, photosynthesis is the source of carbon and energy for all forms of life, whether directly or indirectly. All green plants, green, red, and brown algae, diatoms, dinoflagellates, blue-green algae, purple bacteria, green sulfur bacteria, and purple sulfur bacteria are among the organisms that participate.

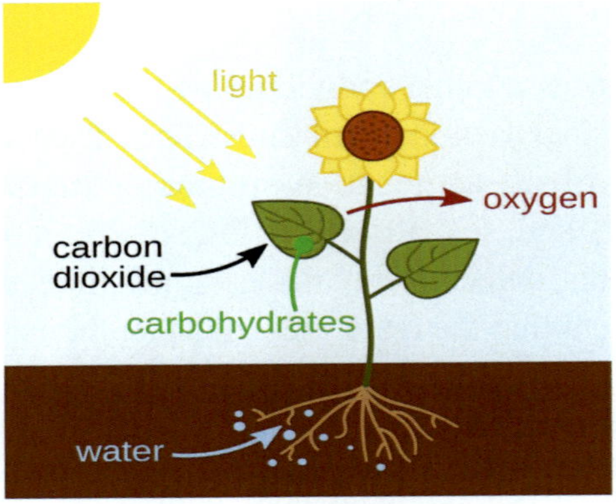


**Figure 56.** A diagram of Photosynthesis .

*Source: Image by Wikimedia Commons*

# Q

## Q Banding

The technique is called chromosome banding since it uses quinacrine dyes (quinacrine mustard or quinacrine dihydrochloride). The technique is very beneficial for displaying heterochromatin regions, particularly the Y chromosomes.

## Quantasomes

Thylakoid membranes of chloroplasts include small molecules with a cross-section of roughly 10 x 18 nm. They're regarded to be photosynthetic activity's structural units.

## Quantum Speciation

On a far smaller time period than generally involved in the development of new species, genetic drift and the founder effect are used to evolve new species within small peripheral populations.

## Quantum Yield

For each light quantum absorbed in photosynthesis, the number of molecules of oxygen produced or carbon dioxide fixed. Quantum yield is directly proportional to the efficiency of photosynthetic chemical reactions.

## Quiescent Centre

At the tip of a plant apical meristem, there is this group of cells that is located and characterized by a much lower rate of mitosis than in surrounding tissue. The center is present in most root and some shoot apices, and may be a site of auxin synthesis.

## Quiescent Culture

When the cell growth in the culture of cells is active, the replication is minimal, and the cells continue to survive adequately. confluent cell cultures may behave in this way.

## Quinacrine

An anti-malarial drug that is a derivative of acridine and is also used in cancer chemotherapy. Quinacrine dyes (quinacrine mustard or quinacrine hydrochloride) are also used as chromosome stains and will fluoresce under ultraviolet light.

# R

## RII Locus

A component of the genome of the bacteriophage T4, made popular because it was subjected to comprehensive study by Seymour Benzer and was the first DNA segment to be subjected to fine structure mapping analysis.

## Race

Subspecific population group distinguishable from other groups or populations within the species by phenotypic characteristics and/or geographic isolation. The degree of differentiation between races varies, and the term is used for convenience. Races may or may not be classified as subspecies, and they are only classified as such when racial distinctions are significant enough to be considered taxonomic.

## Radial Cell

Glial cell that extends from the lumen to the outer surface of the spinal cord and occurs throughout the central nervous system. Radial cells are thought to guide the migrations of other nerve cells to their correct destinations.

## Radiation

Ionizing molecules and electromagnetic radiation, such as gamma rays, X-rays, and light rays in ultraviolet, visible, and infrared wavelengths, are examples of wave energy that is emitted and transported via a medium or space.

## Radiation Chimaera

Ionizing radiation damages the genome of a constituent cell of an organism, resulting in chimera. In the ommatidia of insect eyes, radiation chimeras have been used in clonal analysis to discriminate between lineage-dependent and position-dependent differentiation: clones of cells can be marked by X-irradiation and their fates followed.

## Radiation Sickness

A syndrome that characterizes exposure to high and potentially deadly amounts of radiation, usually above 400 r for the human. Nausea, diarrhea and mental depression are its symptoms; death generally occurs within a few days.

The immediate cause of the disease is not mutation alone, but a generalized loss of capacity for mitotic division of the body's cells. As a result, tissue with high mitotic rates is likely to be the first to show dramatic change.

## Radioactive Contamination

Increased levels of radiation in formerly low background radiation areas due to discharge of radioactive materials. Contamination can occur as a result of fall-out from the atmosphere following an molecular explosion, or from accidental discharge of radioactive gases or liquids.

The severity of such contamination depends greatly on the isotope concerned. Thus, isotopes with long half-lives, intense radioactive discharges and involving elements incorporated into vital biological structures are the most serious.

## Radioactive Decay

In an unstable atom, the disintegration of the nucleus which results in the emission of photons or charged particles such as alpha- or beta-particles. As decay is a random process it is generally expressed as a half-life, being the time taken for 50 percent of a radionuclide to decay.

## Radioactive Isotope (Radioisotope)

An isotope of an element that is innately unstable, which tends to emit ionizing radiation as a result of radioactive decay to a more stable form.

## Radioactivity

When an unstable radioactive isotope randomly disintegrates to produce different atoms with concomitant emission of energetic particles (e.g., electrons) or radiation (e.g., gamma-rays). Radioactivity can be detected in varied ways.

Electrons (beta-particles) can be detected and quantitated in a Geiger counter or in a liquid scintillation counter. Radioactivity can also be located by its action on a photographic emulsion.

## Radioimmunoassay ((RIA)

A method for assaying hormones, drugs, and other substances that are antigenic and present in a very low concentration in biological tissues and fluids. It is dependent on the capacity of the unlabeled antigen to inhibit competitively the binding of radioactively labeled antigen to a specific antibody. Protein hormone concentrations to 1 pg/ml can be measured accurately.

## Random Assortment

At the time of meiotic division, distinct combinations of chromosomes are dispersed among the separate gametes. It stems from the fact that each pair of homologous chromosomes in an individual is made up of one paternal and one maternal chromosome.

Since assortment is random, a single sperm or egg will acquire a paternal chromosome from one homologous pair but a maternal one from some other pair, the origin of the chromosomes being randomly determined.

## Random X-Inactivation

Referring to the fact that In the homogametic sex, such as the human female, the X chromosome that is inactivated in each cell is determined more or less randomly, with the X of paternal origin in some cells and the X of maternal origin in others.

The process isn't completely random, though, because nearby tissue cells often have the same X inactivated, as seen in the tortoiseshell cat, whose coat color betrays the phenotypic expression of two separate X-linked alleles that determine pigment color.

The X chromosomes are randomly inactivated at the whole animal level, but not at the tissue level, therefore the coat color is patchy.

## Rapidly Reassociating DNA

After denaturation, the fraction of DNA that anneals or renatures quickly. The annealing process involves the recovery of double strandedness by specific hydrogen bonding, and the rate at which this occurs in solution is a function of both the frequency of a particular sequence in the mixture and the nature of the sequences involved.

The palindromic sequence material, which can hybridize with itself by doubling over a single-stranded molecule, is the DNA that reassociates the fastest. As a result, it's known as snap back or fold back DNA.

## Rare Bases

In nucleic acids, the nucleoside phosphates, purine and pyrimidine occur infrequently. They include such compounds as 1-methylguanosine, ribothymidine and 5-ribosyluracil which are present in the molecules of transfer RNA.

## Rassenkreis

A phenomenon where a series of graded populations of a species form a continuum, with the extremities of the continuum geographically overlapping but reproductively isolated from one another. The herring gull and lesser black-backed gull in the UK form the overlapping ends of such a continuum stretching around the world in subpolar regions.

## Reading Frame

A specific sequence of nucleotide triplets (codons) employed in translation. It depends on the location of the initiation codon; addition or deletion of one base (or two bases) will alter the reading frame, whereas three base deletions or additions will have no effect on the reading frame.

## Reading Frame Shift

A type of mutation where the addition or deletion of a base within a coding sequence results in a misreading of the code during translation.

This stems from the fact that DNA sequence is read in triplets (codons) from a fixed point of initiation, and any addition or deletion leads to all triplets downstream of the mutation differing from the original. Such base changes frequently result from the actions of intercalating acridine dyes such as acriflavine and acridine orange.

## Read Through

An RNA polymerase's action in continuing to transcribe a DNA sequence beyond a stop codon, or the translational machinery's action in translating a messenger RNA molecule beyond the typical codon for chain termination. For example, the latter type of read through results in transfer RNA, which serves as a nonsense suppressor.

## Reannealing

Renaturation of complementary strands of DNA that have been split by denaturing or melting by particular base pairing. The term is most commonly used to describe the renaturation of previously connected strands prior to melting.

## Reassociation Kinetics

A study of rates of DNA reassociation under a controlled environment, as used in cot curve determination. Following the melting of a DNA sample, the rates of reassociation, as measured by a decrease in absorption at 260 nm, can be used to deduce the proportions of DNA with different Cot values, and hence varying repeating frequencies in the DNA sample.

## Recapitulation

The theory, which is now obsolete that in evolutionary sequence, some or all organisms pass through stages resembling adult forms of their ancestral organisms during development. Such a theory, sometimes summarized in the term "ontogeny recapitulates phylogeny," is widely incorrect, but there are certain cases that support it in a specific, rather than generic application (e.g., the appearance of gill slits in the mammalian embryo, reminiscent of the adult stages of primitive fishes).

## Rec A Protein

In some bacteria, including *E. coli*, the protein which is produced; product of the rec A locus. It has vital functions in DNA repair, especially in SOS repair. The rec A protein is a protease that cleaves the hex A repressor. A cleavage like this activates a number of genes involved in repair.

Because SOS repair is subject to error, its efficiency has a significant impact on the mutation rate.

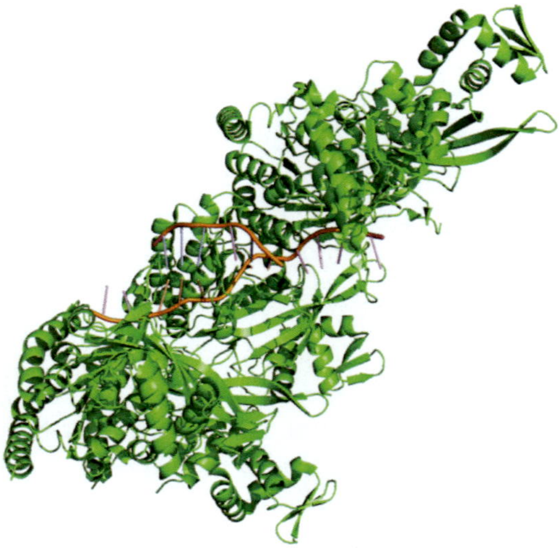

**Figure 57.** Crystal Structure of RecA-DNA Complex.

*Source : Image By Wikimedia Commons*

## Receptor

A molecule or a molecular complex that recognizes and connects with another specific molecule in a certain way (ligand). When the receptor becomes overly complicated, it frequently triggers new molecular processes within the cell.

Bacteriophages have specific receptors on the surfaces of sensitive strains of bacteria, many hormones bind to receptor sites on cell surfaces, and different DNA sequences act as receptors for polymerase molecules, transcription factors, and regulatory molecules involved in the activation or repression of operons at the nucleic acid level.

## Receptor-Mediated Endocytosis

A type of endocytosis which is highly specific for certain important macromolecules found in the extracellular fluid. Uptake is effected through the presence of high-affinity, highly specific protein receptors which become localized in coated pits of the plasma membrane and concentrate the ligand at this site.

The ligand is then internalized after the synthesis of a coated vesicle. The ligand and receptor have different fates. Both, like transferrin, which carries iron into the cell, can be recycled to the cell surface.

## Reciprocal Cross

The pairs of mating wherein males and females of each of two separate genotypes are crossed in such a way that a male of the first genotype, for e.g., AA is crossed with a female aa; in the other mating, a female AA is crossed with a male aa.

This suggests that the outcome of the cross is influenced by sex. For example, reciprocal crossings of horses and donkeys produce mules and hinnies.

## Reciprocal Translocation (Balanced Translocation)

Chromosome aberration which occurs when two non-homologous chromosomes exchange chromosomal segments precisely after chromosome breakage. It's possible that a translocation between homologous chromosomes would go undetected. Reciprocal translocation usually does not change the length or appearance of the chromosomes involved.

## Recombination

Reassortment of genes in progeny as a result of separate chromosome segregation during meiosis and cross-over events that occur at the same time. (2) Reassortment of DNA sequence portions between two DNA molecules.

## Recon

The smallest unit of DNA capable of involvement in recombination. It is a rarely used expression now, for a single base which is closely analogous to a muton. In practical terms, generally, sequences of many bases are involved in a recombination event.

## Redifferentiation

From one state of differentiation to another, when the cellular specialization changes. It's debatable if all cells must dedifferentiate before proceeding towards further specialization or redifferentiation.

The iris cells undergo redifferentiation after the lens is removed from amphibian eyes, which is a classic occurrence of redifferentiation. Iris cells that were previously colored shed their pigment and transform into lens cells, changing shape and generating large amounts of the lens-specific protein crystallin.

## Regeneration

Mostly following injury, the re-growth of lost tissue or organs by an organism. Such growth is very common in plants, mainly because of meristematic tissue's retention in stems, roots and leaves. Hence, vegetative propagation of plants from cuttings of all three tissues is common, the ease of which varies with species, age, etc.

Lower animals regenerate frequently, but as complexity rises, regeneration becomes less visible. Many planarians, coelenterates, and other invertebrates may develop new organisms from tissue explants. Legs and antennae are easily replaced in crustaceans. Complete limbs or tails will regrow in amphibians and reptiles.

## Regression Line

A line superimposed on a graph that shows how much an increase or decrease in one variable affects the increase or decrease in another.

## Regulation

The ability of an organism, generally an embryo, to develop normally in the face of manipulation (e.g., transplantation or removal of cells). As a result, regulation is a compensating system that allows normal growth to continue in the face of an aberrant insult or harm.

## Regulative Egg

During the cleavage stages, the cells of the fertilized egg maintain their equipotentiality. Thus, in regulative eggs, the deletion of a blastomere can be compensated for by others, and a blastomere separated experimentally may give rise to a normal human in certain conditions.

## Regulator Gene

Gene that regulates the transcriptional activity of other genes in the genome, whether they are adjacent by or distant. The regulator gene lacI in the *E. coli* lac operon produces a protein product that represses the operon's operator gene. A single regulator gene in bacterial regulons can control a number of non-adjacent operons. The majority of proteins produced by regulator genes function via allostery.

## Regulon

A cluster of non-adjacent bacterial operons controlled by a single regulatory gene. In *E. coli*, the enzymes involved in arginine metabolism are coded by a family of genes constituting a regulon, which is similar to how many other bacterial processes are arranged genetically.

## Relaxation Complex

Three proteins that work together to convert some *E. coli* plasmids from a tightly supercoiled form to a nicked open circle of DNA. A trigger (e.g., alkali, heat treatment) is required for such nicking, which occurs at only one specific spot in the plasmid DNA.

Two of the proteins in the complex break off during nicking, but the third remains covalently attached to the nicked DNA's 5' end. The nicking of plasmids is required for conjugation.

## Relict (Relic)

A term used to describe a tiny, distinct group of organisms that are descendants of a much larger population. During ice ages, geographical isolation in preferred sites called refugios frequently resulted in the existence of relict populations.

## Renin

A kidney produced enzyme that stimulates the synthesis of angiotensin from liver globulin. Humans with high blood pressure produce more of it.

## Repair Nuclease

Nuclease that identifies a DNA strand with a missing base and cleaves out the damaged nucleotide, along with some adjacent nucleotides, enabling later repair through insertion of an undamaged sequence by DNA polymerase and DNA ligase.

## Replica Plating

A method by which identical patterns of bacterial colonies are transferred from one petri dish to others. A pad of velvet is pressed on to the master plate with growing colonies and then on to a series of sterile agar-coated plates.

By using media of varying compositions in the secondary replica plates, the nutritional requirements of the various colonies can be analyzed simultaneously, and nutritional mutants identified and isolated.

## Replication Origin

Site of origination of replicating fork in a DNA. The site tends to be activated in groups of 2()..80 such origins in a cluster, a so-called replication unit. The precise manner of activation and the recognition site for the replication origin remains unclear. Origins are generally spaced about 30 00..300 000 bases apart.

## Replicon

The length of DNA that serves as a replication unit during DNA synthesis. In prokaryotes, the complete circular genome is a single replicon, while in eukaryotes, a single chromosomal length of DNA contains hundreds of replicons.

DNA replication commences at an initiation site in the replicon's center, and two replicating forks are extended outwards in either direction to form a temporary replication bubble. When all the bubbles have extended outwards to the extremities of the replicon, the replicated regions coalesce to provide entire lengths of replicated DNA.

## Replisome

DNA polymerase and other enzymes form a group to form the Replisome Complex, which starts and extends DNA synthesis at the replicating fork. Despite the fact that replisomes have only been investigated in prokaryotes, they are thought to exist in eukaryotic replicons.

## Repression

The cessation of synthesis or biochemical activity which results from the action of a specific factor, termed the repressor. Thus, catabolite repression involves cessation of enzyme synthesis by inhibition by the product of the synthetic pathway. Specific repressor proteins interacting with the DNA or messenger RNA at specific sites may repress transcription or translation.

## Repulsion

A situation that compares with the coupling configuration. Two non-allelic mutants in a configuration of (a+l+b) are carried by two homologous chromosomes in this situation; in other words, in the trans-configuration, this cis/trans terminology is usually only applied when genes are pseudogenes.

## Rescue

A phenomenon wherein multiple infection of a cell permit recombination of phage plasmids resulting in the retention of viability, either of the cell, the virus or some genetic or biochemical property of either.

## Resistance Transfer Factor (RTF)

Plasmid carrying genes conferring antibiotic resistance. The resistance genes themselves are generally part of transposable elements, and the plasmid also carries DNA coding for its own transfer between bacterial cells.

## Resolving Power

In a light or electron microscope, the ability of a lens to resolve. The wavelength of the illumination source determines the resolving power of a lens, not just its magnifying power. As a result, the stronger the resolving power of the lens and microscope, the shorter the wavelength.

## Respiratory Pigment

A colored compound combining reversibly with oxygen, thereby increasing the oxygen transport capacity of the circulatory system. hemoglobin, haemerythrin, chlorocruorin and hemocyanin are such pigments, some of these occur in erythrocytes in some organisms, others are free in the plasma of different organisms.

## Restriction Mapping

On a DNA fragment, a technique for arranging type II restriction endonuclease cleavage sites. Restriction maps are usually created by digesting a target molecule completely with a variety of restriction endonucleases.

Sequential complete digests allow the relative position of different restriction sites to be determined. Restriction fragments produced in such analyses are then sized by electrophoresis using agarose gel.

In this system, the mobility of DNA fragments within defined size ranges is proportioned to the log10 Mr. The use of known molecular weight DNA standards permits calibration curves to be built from which the size of unknown restriction fragments can be calculated.

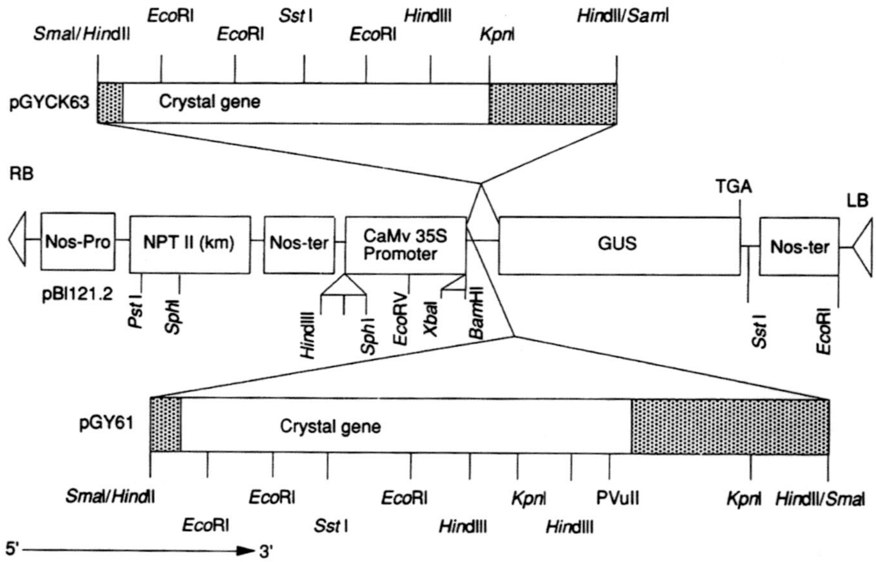


**Figure 58.** Restriction Map Of a Recombinant Plasmid.

*Source : Image By Flickr*

## Reticulocyte

Intermediate between an Erythroblast and an erythrocyte, which is an immature red blood cell. Reticulocytes are so called because they have a slightly stained cytoplasmic reticulum, which is the remnant of the erythroblast's previous endoplasmic reticulum.

Although the human reticulocyte has lost its nucleus, it is still a nucleated cell in other species (e.g., fish, amphibians, reptiles, and birds).

## Retinitis Pigmentosa

Single gene disorder in which clumps of retinal pigment cause progressive visual loss in humans. Both X-linked and autosomal genes are involved in different forms, with an overall prevalence of about one in 5000.

## Retinoblastoma

The malignant tumor of retinoblast - epithelial neurological cells from the retina of the eye - occurring exclusively in children. It is a well-known example of autosomal dominant inheritance of a tumor. Surgery and irradiation can be used to treat retinoblastomas.

## Retroviruses (Leucoviruses)

A group of single-stranded RNA viruses that are characterized by their possessing reverse transcriptase, which transcribes the RNA genome into DNA during the replicative cycle. Later, the transcribed viral DNA is integrated into the host cell DNA and is replicated with it.

The genomes of most retroviruses consist of a set of four genes: gag, which codes for the capsid protein of the virus; pol, which codes for the reverse transcriptase; env, which codes for a protein forming spikes on the outer envelope; one, which is an oncogene. Many retroviruses are oncogenic, by the virtue of possession of oncogenes.

## Reverse Transcriptase

(RNA-dependent DNA polymerase). Enzyme that allows transcription of DNA from an RNA template in retroviruses. Such an enzyme has never been isolated from other sources, but speculation about its possible occurrence continues.

It is largely used experimentally in genetic engineering to allow the formation of complementary DNA (cDNA) from purified RNA. Discovery of this enzyme faulted the widely held central dogma, which stated that the information flow in nature was always from DNA to RNA.

## Ribosomal RNA (rRNA)

The RNA used in the composition of ribosomes. It consists of four different classes in eukaryotic cells, designated as 28S, 18S, 5.8S and 5S in accord with their sedimentation coefficients. The genes for the first three occur as tandem repeats in the nucleolar-organizing centre of the genome, the numbers of such repeats being as high as 450 in Xenopus and 250 in Homo.

Initial transcription products from these repeats consist of a 45S precursor RNA which is processed to yield three distinct RNA molecules. The 5.8S RNA molecule is later hydrogen bonded with the 28S.

## Ribonucleoprotein (RNP)

A complex of RNA and protein. Both the nuclei and cytoplasm of eukaryotic cells contain molecules of RNA that are formed in order to protect RNA from degradation and probably also to assist in its transport and localization within the cell.

## Rifampicin (Rifampin)

The antibiotic inhibitor of RNA synthesis. A derivative of rifamycin SV, rifampicin, particularly inhibits DNA-dependent bacterial and chloroplast, but not RNA polymerase in mammals. It is also reported to inhibit the assembly of DNA and protein into mature virus molecules.

## R Loop

When RNA displaces one of the two strands of native DNA during molecular hybridization it forms this structure. Under certain conditions of formamide concentration and temperature, DNA-RNA hybrids are more stable than native DNA hence, resulting in one of the DNA strands to loop out.

Such structures can be seen under the electron microscope, identifying the section of DNA that codes for a particular RNA. R looping has been used to map intron and exon positions in eukaryotic genes.

## Rolling Circle

A method through which replication of some types of circular DNA molecules (e.g. .. fX174) takes place. One of the two strands of native DNA is nicked and the free 5’ -end attached to a stable support such as an internal cell membrane.

DNA polymerase uses the free 3'-end as a primer and then copies the second (unnicked) strand. During this time the nicked strand unravels and the intact one rolls along it away from the fixed attachment point.

## Root Hair

The outgrowth from an epidermal cell in the root which increases the surface area of the root. They occur in a part of the root behind the area of root extension, known as the piliferous layer. Root hairs make close contact with the soil molecules and are responsible for the absorption of water and solutes.

## Root Nodule

On the roots of certain plants, a growth is developed in response to infection by symbiotic micro-organisms. On leguminous plants, these are caused due to infection with Rhizobium and in these nodules, nitrogen concentration occurs. Nitrogen-fixing nodules also occur on the roots of a few species from other families but, these develop in response to other microorganisms.

# S

## Saccharomyces Cerevisiae

Brewers' or bakers' yeast; a budding yeast extensively used in genetics and cell biology. Mitochondrial mutations which appear as slow-growing petite mutants have been much studied in this organism, as they have the genes controlling mating type.

## Safranin

SA stain which indicates the presence of various structures in plant cell sections, especially those containing suberin and is commonly used in alcoholic solution.

## Salivary Gland Squash

Preparation of salivary gland tissue from larvae of Drosophila and related insects for studies on their polytene chromosomes. In this procedure, cells are squashed in a drop of aceto-orcein and directly examined without prior fixation or sectioning.

## Salmonella typhimurium

Warm-blooded mammals, including humans, harbor a rod-shaped, facultative, gram-negative, motile, non-spore-forming bacterium (where it causes food poisoning). It does not use lactose as a fermentable carbon source, unlike coliform bacteria (e.g., E. coli).

Although it can be grown in standard nutrient media, it is not as widely used as *E. coli* as an experimental organism in biochemistry and genetics.

## Saltatory Replication

Without nuclear division, the increase in the number of gene copies of a specific DNA sequence by laying down numerous strands of DNA in parallel. This increase in the number of copies of ribosomal cistrons occurs in some strains of Drosophila and is evident in the polytene chromosomes where these genes are over-represented (or in other cases, under-represented.)

## Sandhoff Disease

An autosomal recessive gene responsible for invariably lethal disease in humans. Present on chromosome 5, the gene codes fora defective hexosaminidase subunit that causes mental deterioration. Tay-Sachs disease has similar symptoms as well.

## Sarcomere

Surrounded by the Z lines, the functional unit of striated muscle. The sarcomeres lie parallel and in series, thus, giving the muscle its characteristic striped appearance. In each sarcomere, there are thick (myosin) and thin (actin) myofilaments. Active sliding of these filaments across one another shortens the individual sarcomeres resulting in contraction of the muscle cell.

## Sarcoplasmic Reticulum

Specialized for its role in the coordination of the contraction of muscle fibrils, it is the endoplasmic reticulum of muscle cells.

Composed of two components: transverse component (the T system) of vesicles and tubules— its role being to transmit the action potential from the muscle cell membrane into the cell; longitudinal component of anastamosing tubules, which stores calcium.

Upon excitation of the muscle, calcium is released resulting in cytoplasmic free calcium rising from 0.1 to 10 !J.M. This causes muscle contraction.

## Satellite Chromosome

Only by a thin filament of chromatin termed a secondary constriction, a piece of chromosome is connected to the distal end of a chromosome arm. The thin filament may be invisible under low magnification and the small piece looks superficially like a supernumerary chromosome.

## Satellite DNA

The eukaryotic DNA that separates as a distinct fraction after gradient centrifugation. This separation is generally accomplished using isopycnic cesium chloride centrifugation, and is caused by the fact that the satellite band contains material with a different base composition than the main band of genomic DNA (i.e., the satellite band has a higher A + T or G + C content than the main band).

Mitochrondrial DNA, big ribosomal RNA, and families of repeat sequences such as the human Alu family of sequences are examples of DNA that separates as a satellite fraction in various species.

## Saturation Hybridization

When one of the components, in the hybridization of nucleic acids is in great excess, thus ensuring that all complementary sequences in the other sample are hybridized. If the DNA is the molecule in excess and RNA the scarce component, the reaction would be said to be DNA-driven and the DNA is referred to as driver DNA.

## Scanning Electron Microscope

Back-scattered electrons are used in this electron microscope to offer information on the surface topography of solid specimens. Coupled with freeze-etching methods, high-resolution scanning electron microscopy may expose the three-dimensional ultrastructure of cells and tissues in a new and specific way.

## Scattering

Reflection of light rays or electrons due to collisions or interactions, which is random. Light scattering may cause aberrations in spectrophotometry whereas, electron scattering by the specimen is used to provide the image in the electron microscope.

## Schizophrenia

Mental disease with a large genetic component, but the gene, or genes, responsible are unknown. Up to 5 percent of the human population, is affected by it, depending on race and culture, it is characterized by a so-called split mind, often involving severe delusion, paranoia and other types of mental disturbance, but generally interspersed with periods of remission.

## Scission

When a long molecule or a pair of cells, breaks or separates or one piece of tissue breaks from another as in the abscission of leaves (i.e., leaf fall) in deciduous trees.

## Sclereids

Varying in shape but characteristically having a thick secondary wall which is typically lignified, these are plant cells, which form a component of sclerenchyma. Based on their shape and form, they are classified into a number of types. Brachysclereids (stone cells) are shaped like parenchyma cells; occur in the flesh of fruits such as pear.

Macrosclereids are elongated and columnar; found in seed coats of peas and beans. Osteosclereids are bone-shaped; occur in seed coats and leaves. Asterosclereids are branched and somewhat star-shaped. Trichosclereids are highly elongated and hair-like and sometimes branched.

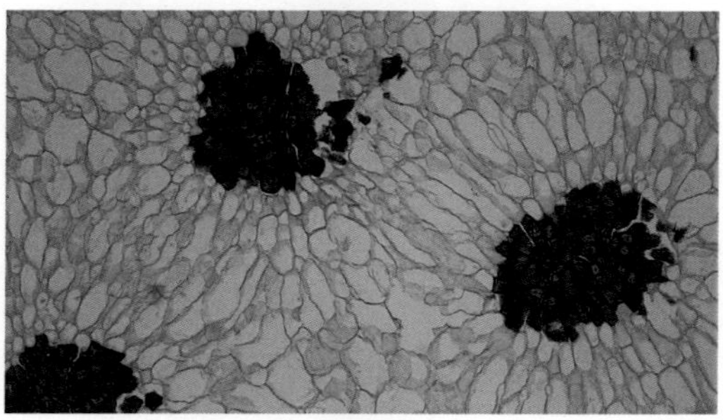

**Figure 59.** Sclereids In Pyrus Pear.

*Source : Image By Wikimedia Commons*

## Sclerenchyma

Plant tissue made up of thick-walled cells, often lignified, with the main function to provide mechanical support. Often, at maturity, the cells are devoid of living contents. Sclerenchyma can be subdivided into fibers which are elongated cells with tapering ends and sclereids which are variously shaped.

## SDS-Polyacrylamide Gel

Modification of polyacrylamide gel electrophoresis in which the proteins' solution to be examined is dialyzed against a solution of the detergent sodium dodecyl sulfate (SDS). The proteins bind to the SDS and all become negatively charged, having similar weight-to-charge ratios.

The electrophoretic separation will occur dependent on the proteins' size and shape, in an electric field. The use of differing concentrations of polyacrylamide gel permits different molecular weight ranges to be analyzed.

## Searle Translocation

Autosomal reciprocal translocation of the X chromosome in the mouse which has the effect of preferential inactivation of the paternal X chromosome in female heterozygotes.

## Secondary Lysosome

In cell cytoplasm the vesicle that results from fusion of a primary lysosome with an endosome (i.e., membrane-bound endocytotic vesicle). Degradation of ingested material occurs in the secondary lysosome.

## Secondary Ray

In the secondary xylem and phloem, a radial line of parenchyma cells arises from a ray initial in the vascular cambium. Rays may be uniseriate (one cell wide) or multiseriate (composed of several cells) when viewed in a tangential longitudinal section.

## Second Messenger

MA molecule which is the carrier of stimulus from the cell surface to the cytoplasm and nucleus following the attachment of a polypeptide hormone to the outer surface of the plasma membrane of a cell. Even though many such hormones do not enter the cell, they transmit their effective stimulus to the cell cytoplasm and nucleus via the second messenger molecule. In most cases, this molecule is cyclic AMP.

## Secretin

When acidified food leaves the stomach, this polypeptide hormone of 27 amino acids is secreted by the duodenal mucosa and stimulates the release of pancreatic juice.

## Secretory Cell

Like many epithelial cells in the pancreas, it is a cell involved in secretion. They have abundant endoplasmic reticulum and numerous secretory vesicles, which release their contents at the cell surface through exocytosis.

## Sectioning

In cell biology and histology, a method in which thin sections are cut from tissues or organisms prior to microscopy. It may be carried out crudely on live plant material, and more precisely using a microtome on fixed material following its embedding in way or resin. For electron microscopy, sections must be ultra-thin and cut with a special microtome using a glass or diamond knife.

## Segregation

A term used at various varied levels in genetics, but most frequently indicating the separation of a pair of homologous chromosomes at meiosis to ensure that only one member of each pair is present in any one gamete.

Segregation of genes refers to Mendel's Law of independent assortment and the fact that alleles on homologous chromosomes are independently transmitted, whereas segregation of chromosomes refers to the separation of whole chromosomes.

Segregation distortion refers to abnormal ratios of segregation which are caused due to lethality or disfunction of gametes carrying particular alleles. In Drosophila, a segregated distorter (SD) mutation has been identified.

## Selection

Most important process in evolution, in which individuals with certain genotypes make a greater or lesser contribution to the population pool of progeny than other individuals do. When the contribution to the pool of progeny is greater, the parental individuals are said to have a selective advantage or 'to be selected for' whereas in the case of lesser contribution, the individuals are said to have a selective disadvantage or 'to be selected against'.

Of a given gene, the selective effect is determined by the probability that the individuals carrying that gene will reproduce and produce viable offspring. Selection pressure may be applied artificially by modifying the culture conditions.

## Selection Differential

The difference between the value of a specific trait in terms of selective advantage when applied to the entire population. and the value of that character in those who become parents to the following generation through selection.

## Selective Neutrality

Individuals in a population having a mutant allele have the same fitness as those in the group carrying the wild-type allele.

## Selective Plating

Technique for isolating recombinant bacteria colonies by mixing two auxotrophic mutants together on a minimal medium agar plate; under these conditions, only colonies indicating chance recombinants will thrive.

## Self-Assembly

The spontaneous aggregation of supramolecular structures such as multienzyme complexes from preformed macromolecules. Protein filaments, ribosomes, membranes and viruses in which the components are non-covalently linked.

The formation of actin filaments, the aggregation of the protein cylinder around the helical RNA core of tobacco mosaic virus, and the maturation of simple bacteriophages are certain examples of self-assembly.

## Self-Replication

The property presents in most cells given a suitable environment and molecules such as DNA and, in rare cases, RNA. The development of self-replication is essential in hypotheses about the origin of life.

## Semipermeable Membrane

Membrane that only allows certain molecules to pass. Biological membranes (e.g., plasma membranes, nuclear membranes and vesicle membranes) are semipermeable; some small molecules such as water are allowed in and out relatively freely. The passage of other molecules such as proteins is impeded.

## Sendai Virus

Enveloped myxo-virus with a single-stranded RNA genome. It caused a widespread infection in laboratory mice, however, now it is now used, in an attenuated form after ultraviolet irradiation for the promotion of experimental cell fusion.

## Senescence

Changes which are deleterious and occur in an organism with age, resulting in a loss of vitality (i.e., the ability of the organism to sustain life in the face of its internal and external environments). All organisms senesce, and the pattern of senescence seems to be characteristic for each species. Even though there are probably various causes of senescence they're not wholly comprehended.

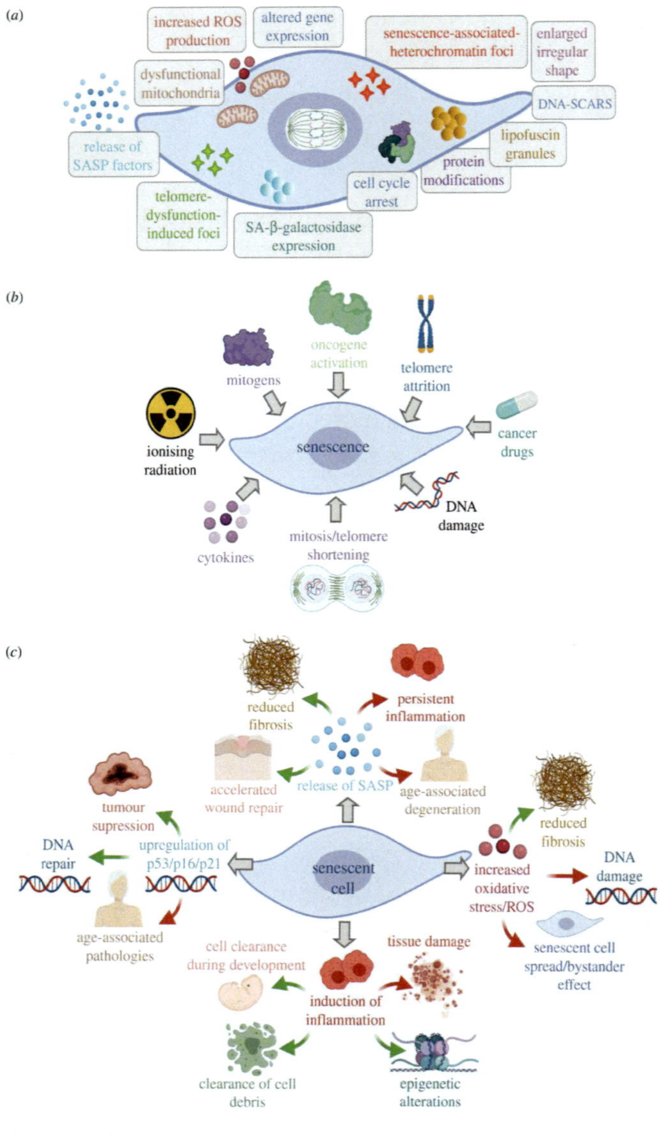


**Figure 60.** The Hallmarks, Causes And Effects Of Cellular Senescence. (A) The Key Features Of A Senescent Cell. (B) Senescence Occurs In Response To Multiple Contingencies Of Life (C) Senescence Plays A Dual Role In Development And Tissue Repair/Regeneration.

*Source : Image By Wikimedia Commons*

## Sensitizing Agent

In a biological system, an agent that increases the damage incurred upon exposure to radiation and potentiates a resultant immune response. Otherwise, potentiating an inflammatory response in a subsequent encounter with an irritant. Thus, an eczematous response to, for example, a detergent is often dramatically increased through prior contact with nickel.

## Septate Junction

A cell junction present in invertebrates' cells which replaces tight junctions in such cells. It differs from tight junctions in the more orderly sequence of the junctional proteins and in that the proteins rather than the membranes themselves form the seal. They serve to hold cells together and form complete septal ribbons around cells, like tight junctions.

**Sericin** A minor gummy protein of silk which forms a covering for the twin filaments of fibroin in each silk Fiber. There are a various different molecular forms of sericin, however, each is composed of more than 30 percent serine.

## Sertoli Cell

Found in the epithelial wall of the seminiferous tubules of the vertebrate testis, it is a specialized cell which provides a vital supportive/nutritional role in the maturation of the spermatozoa (in this respect these cells are often considered homologous to the follicle cell in the ovary).

## Sex

Reproduction which depends on the fusion of haploid gametes produced by meiotic division. (2) The distinction between organisms depending on the type of gametes produced.

If morphologically indistinguishable gametes are produced, they are termed isogametes and the sexes may be termed simply plus and minus, as in mating type in yeast. When an individual produces both male and female gametes it is termed as hermaphrodite, if animal, and monoecious, if plant.

## Sex Chromosomes

The chromosomes that determine sex or contain genes that determine the same. These are dissimilar in the heterogametic sex. The sex chromosomes are the X and Y chromosomes, and the non-sex chromosomes are termed autosomes, in humans.

In Drosophila, although XX is female and XY is male XO is also male, indicating that the number of X chromosomes is the sex determining mechanism, rather than the presence or absence of Y, as in the human.

## Sex-Conditioned Character

A phenotypic character which although is determined by an autosomal gene, is conditioned by an individual's sex. Hence, human baldness is dominant in males but recessive in females; even when females are homozygous, the gene is expressed only to a minor degree.

## Sex Determination

The mechanism that determines sex within any given species. Sex chromosomes represent the commonest method of sex determination; one sex is homogametic and one heterogametic. Often, like in the case of mosquitoes, a single gene containing two alleles is responsible, and one sex is homozygous recessive (female) and the other the heterozygous (male).

## Sex Hormones

Hormones (e.g., estrogens, progestogens, androgens and gonadotrophins) that are secreted by or influence the development of the gonads. Various secondary sexual characteristics are caused due to activity of sex hormones.

## Sex Linkage

A genetic linkage type in which the gene responsible for determining the character concerned is carried on a sex chromosome. Even though in other species it varies depending on the mechanism of sex determination and the types of sex chromosomes present, in humans this is always the X chromosome. This may suggest that the particular phenotypic trait is almost exclusively expressed in the heterogametic sex (e.g., hemophilia in humans). Differences between reciprocal crosses may also arise with sex-linked genes.

## Sex Ratio

In a population, the ratio of males to females. Because individuals of one sex may live longer than those of the other, the statistic is usually adjusted for age. Although close to 1:1 ratios are typical, they are not the rule. One sex, usually the male, is less viable in embryonic life in some species; males may be uncommon or unknown in others.

## Sex Reversal

During an organism's lifetime, the transition from one sex to another. Sex reversal is widespread in some species, such as marine fishes like wrasses, where sex is determined by environmental circumstances. In animals, it is usually caused by pathology or artificial interference, the latter of which is often used for psychiatric purposes.

## Sexual Dimorphism

The existence of morphological distinctions in color, size or other features between sexes of one species, enabling the population to have two distinct types differing by sex. Thus, sexual dimorphism is pronounced in species like the lion and the pheasant, but scarcely apparent in leopard or the partridge.

## Sexual Isolation

Due to differences in courtships or mating behavior, or physical inability to mate, the inability of closely related species or subspecies to produce progeny.

## Sexual Reproduction

The production of offspring which is dependent on the fusion of haploid gametes produced by meiotic division. Contrary to non-sexual reproduction in some lower eukaryotes, such as vegetative growth of yeast, and vegetative propagation in plants.

## Sexual Selection

In a species, the selection of characters ensuring successful male choice, even if such features are of no other value to the species and therefore give no general increase in fitness.

Thus, the tendency for males to be very large or, for example, to have large antlers certainly improves the chances of the individual either defeating other males in combat, or proving more attractive to a female in competition with other males.

Competition between males is termed intrasexual selection and female choice from among competing males is termed epigamic selection.

## Short-Period Interspersion

Gene sequence's arrangement pattern in an organism's genome in which non-repetitious coding sequences are of some 1000 base pairs in length, alternating with short repetitive DNA sequences of about 300 base pairs.

## Shotgun Cloning

When a number of different restriction endonucleases are used to cut up purified genomic DNA prior to cloning of the fragments in bacterial or phage vectors to form a genomic library.

## Shuttle Vector

Plasmid vector replicating in more than one host, for example in both *E. coli* and *Agrobacterium tumefaciens*, or in both *E. coli* and *Neurospora crassa*, enabling the transfer of genes spliced into such plasmids from one cell to another in gene cloning techniques.

## Soluble RNA

Low-molecular-weight RNA of ill-defined composition, but majorly consisting of transfer RNA. The term is sort of a misnomer since all RNA, including high-molecular-weight species, are readily soluble in aqueous media.

## Somatic Hybridization

When somatic cells from two different species (e.g., a human and a mouse cell) fuse normally in tissue culture. A vital technique in gene mapping, since hybrid somatic cells can be obtained with a full complement as in mouse chromosomes, but with only one residual human chromosome. Any human-specific proteins synthesized are clearly coded by genes carried on the single human chromosome.

## Somatic Mutation

Mutation that has occurred in a cell other than a germ cell. Most somatic mutations will have no phenotypic effects as the affected gene may not be expressed in the tissue of which the cell forms a part.

Certain mutations can be detected as a distinct line of cells in the tissue arising from the cell where the mutation originally occurred. Some types of variegation in plants result from somatic mutations, as do certain changes in leaf form or branching pattern.

## Somatostatin

Released by the hypothalamus, a peptide hormone, that inhibits the production of growth hormone (somatotrophin) through the anterior lobe of the pituitary gland. Also inhibits insulin and glucagon production by the pancreas. Its gene has been isolated from humans and cloned in *E. coli*; active somatostatin was recovered from the clone.

## Somite

In vertebrate embryos, a series of blocks of mesoderm, segmentally arranged, lay on either side of the neural tube. Each consists of three regions: myotomes, which develop into striated muscle; sclerotome, which develops into the vertebral column and proximal parts of the ribs; dermatome, which develops into the dermis.

## Sonication

When ultrasound is used to homogenize tissue. The procedure has fallen out of favor in molecular biology as it tends to damage large, long molecules by shearing, however, may be used intentionally to drastically reduce the molecular weight of a nucleic acid solution.

## Sorting Out

In cell and developmental biology, a phenomenon in which cells with distinct surface properties separate, selectively from one another by the differential adhesion of the cell surfaces.

Such differential adhesion can be demonstrated to be tissue-specific, so that a mixed aggregate of cells from differing tissues gradually sorts out into separate aggregates of only one cell type, or aggregates in which one cell type is internal and the other is external in the cell mass.

## SOS Repair

DNA repair process, involving varied distinct enzymes; extremely crucial due to property of error thus generation of mutations. Regulation of SOS repair in *E. coli* is known to depend on the fact that the repair enzymes are coded by genes that have common SOS boxes in their regulatory regions.

These boxes are specifically repressed by a protein called hex A. However, this protein is specifically cleaved and inactivated by arecA protease, which is, in turn, activated by damaged DNA. So damaged DNA is able to induce its own repair.

## Spacer DNA

DNA sequence located between gene-coding sequences that may or may not be transcribed by RNA polymerase. These sequences are often repetitive in the genome and are a conspicuous feature of the sequence of the multiple copies Of large ribosomal RNA (rRNA) genes in higher animals.

Thus, the rRNA is synthesized as a 45S precursor from a long coding sequence arranged in series, and a short section of transcribed spacer DNA lies between the 3'-end of the gene and the 5' -end of the next.

## Species

Ideally defined as a population of interbreeding individuals, which produce viable offspring and are reproductively isolated from other distinct species; it is the basic unit used in the classification of animals and plants.

Although members of the same species often resemble one another closely, the similarity is complicated by sexual dimorphism, polymorphism, and the possible existence of clines, subspecies or varieties.

## Specific Activity

In the same sample, the ratio of radioactive atoms or molecules to non-radioactive ones. Often stated as curies per mole or millimole, or as the amount of radioactive atoms per million nonradioactive atoms.

## Specificity

The selective need on the part of one reactant molecule or cell for its substrate, receptor or macromolecular complex. As a result, antibody is specific for antigen, and an enzyme and its substrate share specificity.

## Spectrophotometer

An instrument used to determine light absorption by solutions of light-absorbing material. Generally, the light beam is monochromatic and of a selected wavelength, measured after passing through the appropriate solution through means of photosensitive detectors.

The intensity of light transmitted is compared with that passed through a control sample in the solvent. Scanning spectrophotometers measure a range of wavelength absorbances, enabling the derivation of an absorbance spectrum for the solution.

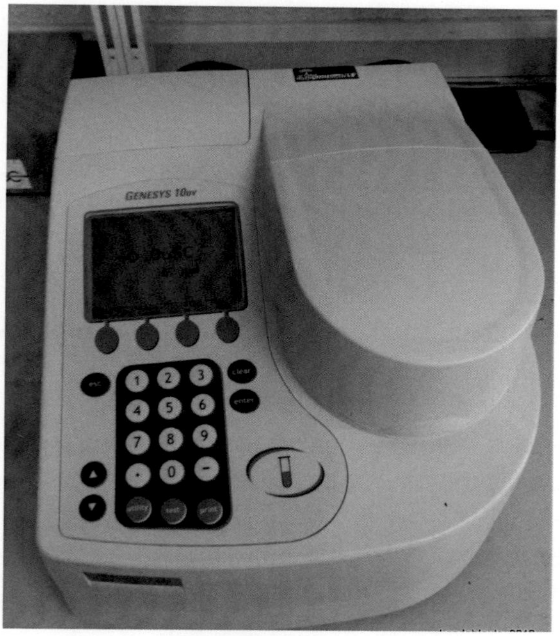

**Figure 61.** Image Showing A Spectrophotometer.

*Source : Image By Flickr*

## Spermatophore

The mass of sperm held together by mucus, that may be deposited inside the female body during copulation (as in some insects) or into water or onto a surface by the male, to be taken up by the female into her cloaca (as in salamanders), after deposition.

## Sperm Bank

With additives such as glycerol, derived from many individuals, it is the store of sperm, normally in semen which is maintained in liquid nitrogen at -196°C for many years. This may be used in artificial insemination. Sperm banks permit the storage of gene pools that might otherwise be lost forever.

## Spindle

At the beginning of nuclear division, a microtubular apparatus that appears in many eukaryotic cells and is responsible for the ordered separation of the chromosomes, which are attached to the spindle fibers through centromeres.

Two types of spindle fibers can be distinguished: the interpolar fiber; stretching continuously from pole to pole of the spindle and the kinetochore fiber or stretching from the pole to the centromere (kinetochore) of an individual chromosome. The mechanism causing the mobility of chromosomes and the spindle fibers to contract is still unclear.

## Spiral Cleavage

An embryonic development found in mollusks and annelids in which the horizontal cell divisions in the blastula result in quartets of cells that are slightly out of line with the quartet above or below, thus, giving the embryo a spiral appearance. Complex genetic influences, including strictly maternal influences, determine the direction of the spiral.

## Spiral Thickening

In tracheids and xylem vessels, the pattern of wall formation where the thickening is laid down in spiral bands is usually secondary to other forms of thickening in these cell walls.

## Spirochaetes

A unicellular bacteria with flexible, elongated cells arranged in a helix around an axial fibrillar filament. The cells are enclosed within an outer flexible, close-fitting envelope. While some spirochaetes are free-living, others are parasitic or pathogenic. Syphilis in humans is caused by the spirochaete *Treponema pallidum*.

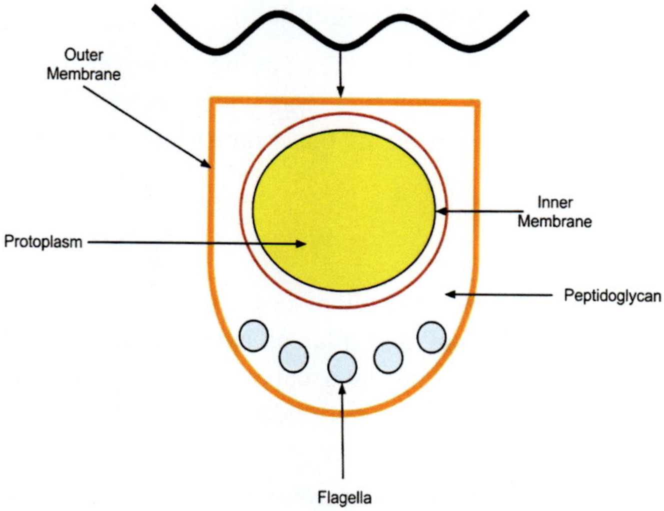


**Figure 62**. Spirochete Cross Section.

*Source : Image By Wikimedia Commons*

## Spleen Colony-Forming Cell

A bone marrow cell injected into the circulation of a mouse following its irradiation. This 'rescue' the mouse through replacement of its own killed bone marrow cells, many growing to form colonies in the spleen. Examination of such colonies has been utilized to shed light on the proliferative cells of the marrow and how they differentiate into lymphocytes, red cells and macrophages.

## Splice Junction

A short sequence residing at the end of an intron that facilitates intron excision and exon splicing. Such sequences at either end of an intron are termed donor junctions and acceptor junctions, being at the 5'-end and 3'-end, respectively, and UsnRNA contains a sequence complementary to both.

## Split Gene

A gene consisting of both exon coding sequences and intron non-coding sequences interspersed. Most eukaryotic genes are split, as are the genes of some animal viruses. Prokaryotes genes lack introns, as do all but a few mitochondrial genes.

## Spontaneous Generation (Abiogenesis)

Doctrine that stated spontaneous arising of living organisms from non-living material. Until the late 18th century, it was widely believed in. Only when sterilization procedures were used did the contrary become evident, as in the work of Louis Pasteur.

## Spore

Reproduction units are commonly microscopic in size. Spores are produced via bacteria, protozoa, fungi and plants, and are adapted for specific functions. Spores might also act as a capability of asexual reproduction for fast vegetative spread. Others may additionally be characterized as a resistant, resting stage in the life cycle.

In plants, spores are produced through the sporophyte after meiosis and are then haploid. Sporophyte after meiosis germinates to produce the gametophyte. The term spore is frequently prefixed to provide an indication of its function (e.g., endospore, megaspore, microspore, zoospore, zygospore).

## Stab Culture

The culture of the microorganism is made by stabbing a long needle, carrying some of the microorganism from some other culture, deep into an elongated check tube containing solid nutrient agar. It represents an easy way of culturing notably non-fastidious anaerobic bacteria given that the oxygen concentration deep in the agar is low.

## Staining

The process is used in temporary or permanent preparation for light or electron microscopy in order to assist in visualizing an element or supplying larger contrast. For light microscopy, many stains are colored dyes with specificity for unique structures, and some, such as the Feulgen stain for DNA, are quantitative.

Cells or cell structures that stain with primary dyes are stated to be basophilic, these staining with acidic dyes, are acidophilic. Vital staining entails staining of structures within dwelling cells. Some stains can be used to localize certain biochemical reactions in cells.

## Starch

Polysaccharides are used as a storage compound in plants; an a-o-glucose polymer. Amylopectin, which consists of 75 percent of starch, is primarily based on 1,4-linked glucose molecules with numerous 1,6-links. Amylose, comprising 25 percent of starch, is based totally on 1,4-linked glucose molecules except for the 1,6-links.

# T

## Tailing

Attachment of a string of adenine nucleotides to the 3'-end of many hnRNA and mRNA molecules after initial synthesis, however prior to transport from the nucleus to the cytoplasm. Tailing is believed to be a machine for identifying the age of a mRNA molecule and so as to make sure its selective degradation at a determined time, consequent on the gradual diminution in the dimension of the tail.

## Tandem Duplication

The scenario in which a section of a chromosome is duplicated and more copies or copies are inserted in line in the chromosome. Thus, the genes for larger ribosomal RNAs are tandemly duplicated in eukaryotes.

## Tannin

Plant phenol by-product that is used commercially due to the fact of its attachment to animal hide to form leather. Tannins are extensively distributed in plants, giving an astringent taste and pigmented coloration of orange, brown and yellow. They are ample in leaves, seed coats and bark, and frequently occupy specialized enlarged cells termed tannin cells.

## T Antigen

Antigen was discovered on normal human erythrocytes but is solely uncovered through treating them with neuraminidase. An ordinary component of serum is anti-T anti-body which can motivate agglutination of erythrocytes contaminated with microorganisms considering the latter produces small portions of neuraminidase.

## Target Theory

The speculation was once developed to account for the mutagenic outcomes of ionizing radiation on cells. It is postulated that solely a small target exists in the cell and that one or more ionizing activities or 'hits" are fundamental within the target area for radiation to produce successfully a mutation. This target is now believed as being the DNA in the chromosomes.

## Taurine

A rare amino acid includes a sulphonic acid team that takes place in the bile salt taurocholate and has an effect on some neurons of the nervous system.

## Taxonomy

Study of the ideas and strategies for classification of the version and variety of living organisms. The term is frequently used synonymously with the time period systematics, even though strictly the latter is involved with the description of residing organisms and the relationships that exist between them.

## Tay-Sachs Disease

An inherited disease, caused by homozygous inheritance of a recessive gene carried on chromosome 15, a gene which codes for the alpha-subunit of hexosaminidase A. In affected individuals, gangliosides accumulate in the brain, and mental retardation and paralysis are accompanied by early death.

The disorder is found most often amongst Askhenazki Jews, of whom some 3 percent are heterozygous for the allele.

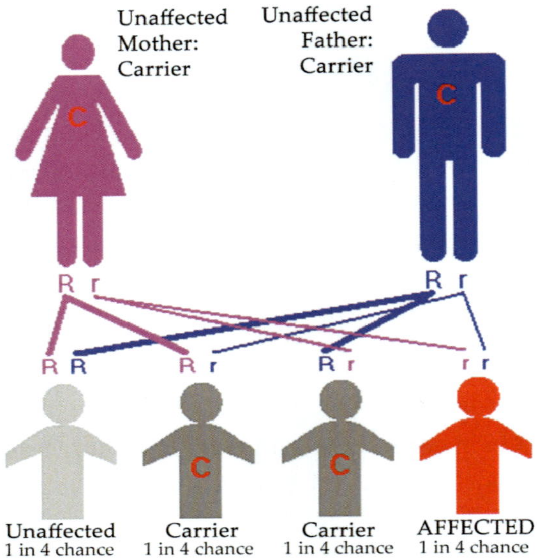


**Figure 63.** Auto-recessive Tay Sachs Disease.

*Source : Image By Wikimedia Commons*

## Teichoic Acids

Molecules are found particularly in bacterial cell membranes (as glycerol teichoic acids) or cell walls of gram-positive microorganisms (as ribitol teichoic acid). They are linear polymers of glycerol phosphate or ribitol phosphate, respectively, with phosphodiester linkages.

## Telegony

The speculation is that the offspring of one mating can be influenced by the genes of a male that had earlier mated with their mother. The idea is no longer accepted in species where more than one mating with special males may also result in litters of reality combined parentage.

## Teleonomy

The idea that the presence of a structure or function in an organism can be assumed to mean an earlier evolutionary gain in the possession of such a shape or function.

## Temperature-Sensitive Mutation (TS Mutation)

The mutation indicates phenotypic expression solely within a constrained temperature range. This situation frequently consequences of the protein coded via the mutant allele being unstable above a certain temperature. Below this crucial temperature, the person carrying the mutant allele is phenotypically normal.

## Template

A molecular structure that is effective in dictating the shape of any other molecule copied from it. This different molecule will regularly be an anti-template, and can in turn dictate the shape of a new template molecule. Thus, a single strand of DNA or RNA acts as a template strand for the elaboration of the complementary antitemplate strand.

## Teratocarcinoma

Malignant and transplantable structure of a teratoma. It can also comprise a weird variety of tissue kinds without morphogenesis or organization, however importantly consists of stem cells, the embryonal carcinoma cells, which may also be isolated and grown as a cell line.

Such cells can be used to learn about differentiation *in vitro* and, interestingly, will make a contribution to regular improvement if transplanted again to a normal blastocyst.

## Teratoma

A tumor that generally consists of derivatives of the three embryonic germ layers (ectoderm, endoderm and mesoderm) (e.g., skin, muscle, gut, and also embryonal carcinoma cells). It is believed to originate each from primordial germ cells and the disorganized boom of early embryos. The word is derived from the Greek root teraton which means a monster.

## Terminalization

Process in which chiasmata have a tendency to move in the direction of the ends of chromosomes in the course of meiosis.

At one time it used to be assumed that terminalization approved chromosomes to disengage from one another and entire segregation, however, it is now clear that such terminalization does not generally go to completion. Instead, the DNA is reduced and rejoined to allow the separate chromosomes to cross apart.

## Terminal Redundancy

Repetition of the identical DNA sequence at either end of a long DNA molecule. Some bacteriophages have DNA that is terminally redundant and can be circularized through the elimination of a few bases from each end.

## Terminal Transferase

The enzyme that will synthesize homopolymer single-stranded tails at the 3'-ends of DNA molecules. It is used in gene manipulation technology, most often to add poly A tails to one DNA duplex and poly T tails to another, for that reason allowing the formation of a hybrid molecule of cloned circular DNA. Recombinant plasmids for cloning in *E. coli* are often built via such homopolymer tailing.

## Terpenes

ipids are made up of multiples of unsaturated hydrocarbon isoprene. Many terpenes are found in plants, the place where they are accountable for characteristic odors and flavors. In the strict sense, terpenes have the empirical formula $C_{10}H_{16}$ and are volatile fragrant hydrocarbons.

In a broader sense, the term consists of sesquiterpenes ($C_{15}H_{24}$), diterpenes ($C_{20}H_{32}$) and higher polymers. The term can additionally be extended to cover a variety of various oxygen-containing compounds derived from terpene hydrocarbons. The term terpenoid is occasionally used to describe all of these compounds.

## Test Cross

Mating between two individuals, one of known and one of unknown genotype, in order to ascertain the genotype of the tested parent. Normally the tester parent is chosen to carry solely recessive genes so that any dominant alleles carried via the examined parent will be revealed.

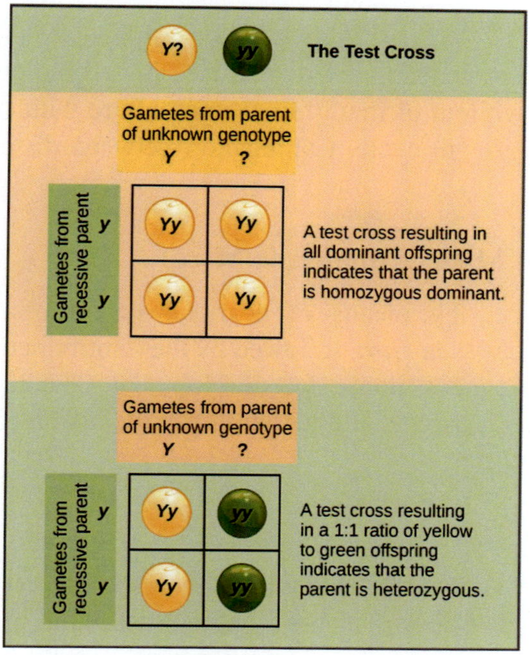


**Figure 64.** Image Showing Test Cross.

*Source : Image By Wikimedia Commons*

## Testosterone

Principal and most effective androgen. Secreted basically through the Leydig cells. In the circulation, nearly all of the testosterone is either a protein that also binds different steroids, sex hormone-binding globulin (SHBG), or albumin.

Testosterone is responsible for developmentthroughout embryogenesis and at puberty, and impart the male secondary sexual characteristics. Androgens, together with testosterone, additionally have an anabolic effect. ensuing in the increased rate of growth and characteristics in males at puberty.

## Test-Tube Baby

Production of a baby following *in vitro* fertilization, after which the embryo is transplanted into the uterus of its biological mother Or a surrogate mother to allow implantation and regular gestation.

## Tetrahymena Thermophill

Species of ciliate protozoan are broadly used in cell biology and genetics. It can be grown without difficulty in axenic culture with fast cell division in its vegetative growth cycle. Conjugation between distinct lines is also common.

## Tetra-parental Mice (Allophenic Mice)

Mice are produced by means of artificial fusion of cells from two extraordinary blastulas *in vitro*, followed by the reintroduction of the hybrid embryo into a pregnant mouse for development. The resulting progeny have consequently 4 parents alternatively than two and incorporate cells of distinct genetic origins.

## Thalidomide

The compound was used as a tranquilizer in the early 1960s. Although no undesirable side-effects had been obvious in adults, it proved to be a teratogen, inducing phocomelia in some infants born to females who had taken the drug throughout pregnancy.

## Thallus

Form of the plant body, discovered in many small plant's life cycle and in the gametophyte stage of higher plants, consisting only of parenchyma tissue with no vascular bundles and no differentiation into leaves, stems or roots.

## Thermal Denaturation

Destruction or partial destruction of the structural conformation of a molecule through heating. Although each protein and nucleic acid are simply denatured by increased temperature, the time period is most frequently used for nucleic acids, in which particular strand separation or melting occurs at accelerated temperatures, and renaturation or annealing might also comply with when temperatures are lowered. Such denaturation and renaturation with the aid of temperature modulation is employed in the determination of cot plots and also in nucleic acid hybridization technology.

## Thermogenesis

Production of internal heat; is mainly utilized in techniques in hibernating mammals and different cold-adapted mammals (including the newborn) via which the mitochondria of brown fat serve as ATP generator for the production of heat energy.

## Thrombin

A proteolytic enzyme that is capable to result in the conversion of fibrinogen to fibrin and so to enhance blood clotting. Thrombin is not normally located in circulating blood, however as an alternative is repressed through its inactive precursor prothrombin. The conversion of prothrombin to thrombin requires the presence of blood platelets, calcium ions, and thromboplastin.

## Thromboplastin

Protein is required, alongside other factors, for the conversion of prothrombin to thrombin during blood clotting. These proteins are located in the blood that has been exposed to injured tissue and can be recovered from intact lung, placenta, brain and different tissues.

## Thylakoid

Flattened sac bounded via a membrane that forms ~ the lamellar grana of the photosynthetic membrane in chloroplasts. Thylakoids additionally form the photosynthetic systems of photosynthetic bacteria. Chlorophylls and other pigments of the photosynthetic pathway are positioned in the thylakoid membranes.

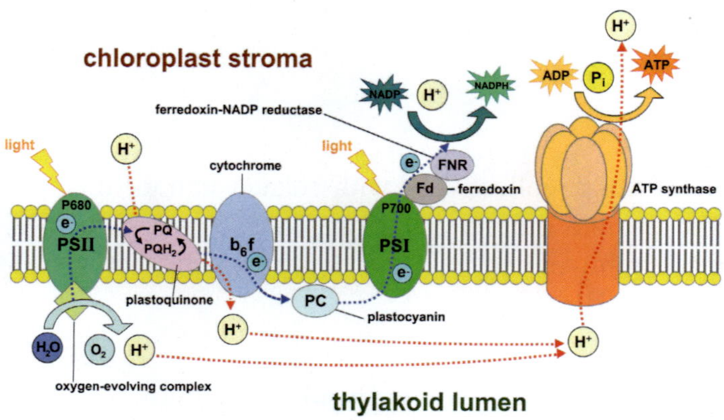


**Figure 65.** Light-Dependent Reactions Of Photosynthesis In The Thylakoid Membrane Of Plant Cells.

*Source : Image By Wikimedia Commons*

## Tight Junction (Zonula Occludens)

Continuous, band-like, intercellular junction around the lateral surface of epithelial cells that serves to seal epithelia towards leakage. The Outer leaflets of the plasma membrane of adjoining cells appear to have fused, thereby apart from completely any intercellular space. Tight junctions are not often viewed in tissues other than the epithelium.

## Time-Lapse Photography

The method used in the production of videos of transferring cells makes use of a photomicroscope and a camera which takes exposures at set intervals of, say, one body per minute. By projecting the film at a greater fast speed, positive factors of over a thousand instances the lapse of real-time are effortlessly obtained, allowing the designated study of the mechanisms of cell movement.

## Ti Plasmid

Large (150-230 kilobases) plasmid carried by *Agrobacterium tumefaciens* strains that induce crown gall in plants. A small phase of this plasmid, the T-DNA, is transferred to and stably built into the host plant's nuclear genome and is accountable for the oncogenic transformation of the plant cells.

One of the attribute elements of the crown Gail tumor cells is the production of opines. The genetic records for the synthesis of opines reside on the Ti plasmid and are transferred to the plant cell genome as part of the T-DNA.

## Tissue

Aggregation of cells, collectively with related extracellular matrices, is specialized to operate differentiated functions. Some tissues can also be made up considerably of one cell type, or of many distinctive cell types.

## Tissue Culture

Maintenance of cells in vitro, in suspension subculture in a liquid medium or on a surface such as glass, plastic or agar furnished with a liquid medium. Primary cultures are those grown from an explant of tissue taken at once from a living organism. Such cultures have a restrained survival time, even though some cells may also live on and grow to be established in culture, dividing regularly and indefinitely.

Such cells are usually transformed, often have chromosomal abnormalities, and develop in a manner similar to cancer cells. They are frequently assumed to harbor integrated viruses.

## Titre

The measure of the number of antibodies present in serum. Normally the titer is decided by deciding the simplest dilution at which the antibody will proceed to agglutinate antigen, or the titer is expressed as the reciprocal of this dilution.

## Tolerance

State of specific, immunological non-reactivity to a given antigen which in different situations would elicit an immune response. Tolerance to an antigen can also be obtained if the animal is exposed to it early in fetal existence or, as an adult, is exposed to very excessive or very low doses of the antigen.

## Tonofilament

Pre-keratin filaments are anchored on the cytoplasmic side of spot desmosomes and extend as a network through the cytoplasm of epithelial cells.

Tonofilament networks in adjoining cells are consequently circuitously linked through the spot desmosomes and represent a tonofilament array running throughout an epithelium. Such arrays can have a vital function in any folding morphogenesis displayed with the aid of the epithelium.

## Tonoplast

The membrane that bonds the vacuoles of plant cells. It consists of a single unit membrane with little mechanical power and so bulges in response to hydrostatic strain from inside or outside, therefore assisting to keep turgor pressure inside the cell.

## Tortoiseshell Cat (Calico Cat)

Female cat in which black and orange patches of fur make a mosaic pattern in the coat. This mottled coloration results from random inactivation of one or other X chromosomes. Each group of cells of one pigment coloration represents a cell clone in which a specific allele of the pigment gene, carried on the X chromosome is expressed.

## Totipotent

Describing a cell (or nucleus) that retains entire potential, so that it is capable to differentiate into all the cell types represented in an adult organism (or cell), or providing rise to cells of all types on transplantation into suitable cytoplasm (or a nucleus). Although the zygote is totipotent, nearly all differentiated cells are not.

## Toxin

A substance that is toxic to cells or organisms and is the product of any other organism. Microorganisms, fungi, some plant tissues and animal secretions are toxins.

## Tracheid

Form of a tracheary element that is elongated and typically pointed at each end. Tracheids are imperforate and the main wall is usually non-stop with pit-closing membranes existing in the bordered pits. They are a characteristic element of the xylem and are involved with water conduction and mechanical support.

All groups of vascular plants have tracheids, however, in most pteridophytes and gymnosperms the tracheary elements consist solely of tracheids.

**Figure 66.** Stem-Tracheid.

*Source : Image By Wikimedia Commons*

## Transamination

Biochemical transfer of an amino group from an amino acid to a keto acid, ensuing in the production of novel amino acid and keto acid molecules. Pyridoxal phosphate is a coenzyme for all transaminases.

## Transcription

RNA synthesis from a DNA template includes a rewriting of the code besides a change in its language. Transcription is catalyzed via RNA polymerase and is solely nuclear in eukaryotic cells (apart from mitochondrial and chloroplast transcription).

Distinct species of RNA polymerase are accountable for the transcription of ribosomal RNA genes, transfer RNA and SS RNA genes, and genes coding for the protein.

## Transcription Control

Control of gene expression at the level of transcription, as distinct from post-transcriptional or posttranslational regulation.

Transcriptional control of gene resynthesis may also be positive or negative. The Lac operon of *E. coli* is negatively controlled and has to be derepressed through the inducer for transcription to occur. Many different genes in prokaryotes and eukaryotes are recognized to be positively controlled.

## Transcription Factor

Protein is essential in effecting transcription of a particular gene sequence. One such factor recognized as TFIIIA (transcription factor/ RNA polymerase III, A), has been discovered to bind to the central function of the SS RNA coding sequence in Xenopus and to facilitate significant transcription of the sequence by means of the type III RNA polymerase. No doubt many different transcription factors exist.

## Transfer Cell

Plant cells with several mitochondria. considerable endoplasmic reticulum. and especially involuted plasma membrane. Transfer cells are ample in conditions where high rates of solute drift occur. As in xylem and phloem loading and unloading. They can also additionally be concerned with glandular and secretory plant tissues.

## Transgenic Organism

An organism that has to turn out to be modified following the introduction of novel DNA into its genome. Such introduction is most often carried out via the integration of cloned DNA sequences following their injection into the fertilized egg. Transgenic animals produced through this technology consist of mice, *Drosophila, Xenopus* and some fish species

## Translational Control

Regulation of gene expression at the degree of translation. Such control is in particular a count of modulating the charge of translation as an alternative to editing the message itself, however, the timing of gene expression may also be modulated at this level by sequestering mRNA for translation long after its preliminary transcription.

# U

## Ubiquinone (Coenzyme Q)

Quinone with an extended isoprenoid tail; the amount of isoprene units depends on the species during which the quinone occurs. In mammals, there are 10 such units (hence ubiquinone-10 or ubiquinone 10).

Ubiquinone may be a membrane soluble component of electron transport chains which mediate hydrogen transfer between membrane-bound dehydrogenases which contain flavoprotein and iron-sulfur proteins, and membrane-bound cytochrome complexes containing b-type cytochrome.

The coenzyme alternates between the quinone and quinol forms during hydrogen transfer and therefore the radical semiquinone is perhaps an intermediate within the reaction.

## Ubiquitin

Acidic protein is widely, and maybe universally, distributed among eukaryotes. It binds strongly to chromatin and should have a structural role in chromatin conformation. Binding specifically to histone H2A by a covalent linkage, the complex has come to be referred to as protein A24.

## Ui RNA

Species of small nuclear RNA (snRNA), of 165 bases long, that's complementary to the donor and acceptor splice junctions and is therefore believed to be involved within the excision of introns and splicing of exons. The UI RNA is related to a selected protein that is involved in intron excision, and together they form the UI ribonucleoprotein particle called either the UI particle or snRNP.

## Ultracentrifuge

Instrument during which tissue homogenates or solutions are rotated at high speed and thereby subjected to intense force, sometimes within the region of several hundred-thousand-fold gravity. Subcellular molecules (cell membranes, mitochondria, ribosomes) and macromolecules (mt-RNA, rRNA, mt-DNA) sediment rapidly consistent with their size, shape, density and relative molecular mass.

There are basically two sorts of ultracentrifuge: preparative and analytical. Preparative ultracentrifugation is employed to purify components for further use, and therefore the centrifuge is solely designed to get large centrifugal forces for a group period of your time.

## Ultrastructure (Fine Structure)

A detailed structure of the cell was revealed by microscopy. The microscope features a higher resolution than is feasible with the sunshine microscope because the wavelength of electrons (0.005 nm) is shorter than that of sunshine (550 nm).

Since the penetrating power of electrons is low, only ultrathin sections of fixed and thus non-living tissue are often examined by transmission microscopy unless a specialized, very high voltage, transmission microscope is employed.

## Ultraviolet Absorption Curve

A plot that shows the connection between different wavelengths of sunshine within the ultraviolet range and therefore the percentage absorbance by an answer of biological molecules. Such a curve helps to differentiate nucleic acids from proteins; the previous absorb maximally at about 260 nm and therefore the latter at about 280 nm.

## Ultraviolet Irradiation

Irradiation with ultraviolet may induce melanin formation in low doses, but is additionally a mutagen and, in high doses, lethal to living cells. it's widely used for the upkeep of sterility in cabinets used for tissue culture, but won't penetrate glass or plastic efficiently. Its mutagenic effect results from its induction of pyrimidine dimers in DNA.

## Ultraviolet Microscope

A microscope during which ultraviolet is generated from the sunshine source, usually a high-pressure mercury vapor lamp, and which is given excitation and barrier filters in order that ultraviolet can excite fluorescent molecules within the specimen, harmful ultraviolet are often filtered out, and fluorescent light only allowed to undergo the eyepiece.

**Figure 67.** Ultraviolet Microscope-1904.

*Source : Image By Wikimedia Commons*

## Uncondensed Chromatin

Chromatin isn't highly compacted and is, therefore, more likely to be transcriptionally active; there's a robust correlation between gene activity and chromatin de-condensation. Both the puffs and inter-band of polytene chromosomes are uncondensed and are sites of intense and moderate transcriptional activity, respectively.

## Uncoupling Agent

Weak acid that's, to some extent, lipid-soluble. It can therefore mediate the movement of protons across membranes, then uncouple electron transport from phosphorylation of ADP.

## Unnamed Hypothesis

The concept is that an unduplicated chromosome, that's one before the S phase of DNA replication, features a single molecule of DNA duplex running from end to finish.

The concept is now generally accepted, although it's clear that the molecule is very twisted upon itself both by reason of packing around nucleosomes, packing of nucleosomes within the 30-nm fibers of chromatin, and further packing of those fibers relative to the chromosome axis.

## Uniparental Inheritance

Type of inheritance during which all the offspring from one mating resemble just one of the 2 parents in one among a variety of characteristics. Such a phenomenon is usually an example of maternal inheritance and is explained by the presence of cytoplasmic determinants within the egg.

## Unique DNA

The sequence of nucleotides presents just one occasion per haploid genome. the good majority of prokaryotic DNA and most protein-coding genes in eukaryotes are unique in this respect.

## Universal Donor

A term used to describe human blood type O donors within the youth of transfusion . All blood is potentially incompatible due to the presence of a different types of erythrocyte antigens and iso-agglutinins and must be crossmatched with the patient's serum before transfusion.

## Unscheduled DNA Synthesis

The synthesis occurs sometimes aside from the S phase of the cell cycle. Most of such synthesis is in DNA repair, although the synthesis of mitochondrial DNA within the cytoplasm is additionally continuous.

## Un-Transcribed Spacer

DNA sequence that lies between two coding sequences and isn't transcribed by RNA polymerase. Although such sequences do exist, some identified within the past (e.g., the spacers between the multiple copies of rRNA genes) are now known to be transcribed.

## Urea ($H_2NCONH_2$)

Water-soluble organic compound; the most nitrogenous body waste of aminoalkanoic acid catabolism in mammals and lots of other animals. it's produced within the liver via the ornithine cycle.

## Uridine Diphosphate Galactose (UDP Gal)

Quinone with a prolonged isoprenoid tail; the quantity of isoprene units relies upon the species at some stage in which the quinone occurs. In mammals, there are 10 such units (subsequently ubiquinone-10 or ubiquinone 10).

Ubiquinone can be a membrane soluble aspect of electron transport chains that mediate hydrogen switch among membrane-sure dehydrogenases which include flavoprotein and iron-sulfur proteins, and membrane-sure cytochrome complexes containing b- cytochrome.

The coenzyme alternates among the quinone and quinol system at some stage in the hydrogen switch and consequently, the unconventional semiquinone is possibly an intermediate in the response.

# V

## Vaccination

Production of protective, energetic immunity through management of a vaccine. The time period become firstly coined by Edward Jenner who used poxvirus Officinal (cowpox virus) to vaccinate individuals to defend them against smallpox (variola).

## Vaccinia Virus

Cowpox virus is used for vaccination against smallpox. It is a poxvirus with a double-stranded DNA genome.

## Vacuole

Fluid-stuffed membrane-sure compartment inside a molecular. Lysosomes are examples of vacuoles in animal cells, as are the endocytotic and pinocytotic vacuoles fashioned at some stage in different processes.

Contractile vacuoles are utilized by many Protozoa as a way of controlling osmotic stress and water content. Most plant and fungal cells have a big conspicuous vacuole bounded through the tonoplast. The vacuole is crucial in biochemical homeostasis and for the restoration of turgor stress.

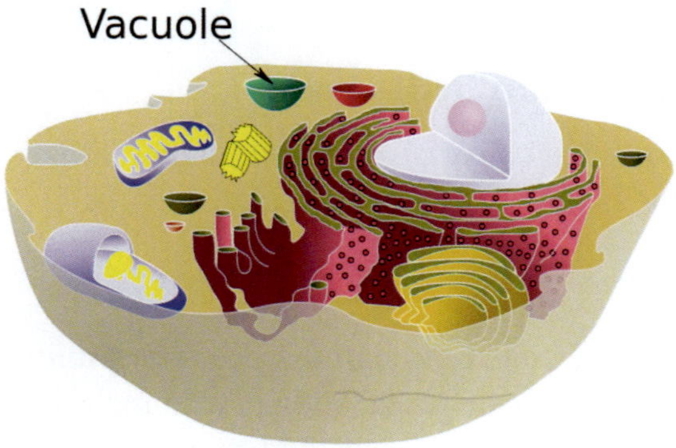


**Figure 68.** Diagram Of An Animal Cell With Only The Vacuole Labeled.

*Source : Image By Wikimedia Commons*

## Van Der Waals Forces

Weak, short-lived forces of attraction among atoms or molecules, mainly among non-polar molecules and consequently of a hydrophobic nature. These forces result from interactions among various dipoles.

## Variegation

Irregularity of pigmentation, normally ensuing in vegetation because of differential chloroplast distribution in a chimera or to viral infection, and in animals because of such elements as nearby physiological situations or X chromosome inactivation.

## Variety

Group of organisms forming a subset inside a species, with one or greater distinct traits that breed true and are consequently determined through genes carried inside the homozygous condition. Often a selection is best maintained as a discrete institution through breeding and artificial selection. There isn't any clear distinction between a strain and a selection, however, the variations are normally extra in the latter.

## Vascular Tissue

(1) In animals, all tissues concerned with the movement of blood, inclusive of the heart, arteries, veins and capillary networks. (2) In vegetation, the transporting tissues are the vascular bundles (i.e., phloem, xylem and related cambium).

## Vasopressin (Antidiuretic Hormone)

A peptide hormone that constricts arterioles and promotes the restoration of water through the kidney tubules. It is secreted by the hypothalamus.

## Vasotocin

Peptide hormone is secreted through the posterior pituitary gland of lower invertebrates and through the pineal body in mammals. It will increase pores and skin permeability in amphibians and regulates bladder characteristics and is structurally much like vasopressin.

## Vector

(1) Organisms liable for parasite switch from one host to another.

(2) Plasmid or different self-replicating DNA molecule that transfers DNA among cells in nature or in recombinant DNA technology. In the latter connection, it can be known as a cloning vector or cloning vehicle.

## Vegetative Petite

Petite mutant of yeast cells wherein a genetic mutation is inside the mitochondrial genome. When this sort of mutant is from a nuclear mutant, it is a segregational petite.

## Vegetative Pole (Vegetal Pole)

Region of the eggs of true animals. The eggs of higher animals are spherical; however, many have an asymmetrical distribution of pigment, molecular organelles, etc. The decreased maximum pole of the egg is the vegetative pole. In a few animals, the cells that form in those areas supply upward thrust to precise germ layers or embyronic tissues.

Certain situations or chemical compounds purpose an overproduction of the embryonic tissues usually produced on the vegetative pole; such atypical embryos are stated to be 'vegetalised'. Other chemical compounds or situations can produce 'animalized' embryos.

## Vegetative Propagation

Production of new organisms without a sexual procedure as budding in Hydra, simple fission in a microorganism or yeast cells, and propagation through cuttings, bulbs, corms, etc. in green plants.

## Vernalization

The procedure through which seeds are exposed to low temperatures earlier than or at some stage in germination as a way of making sure flowering and fruiting in the season of sowing. It is used with cereal plants to obtain a spring sowing accompanied by autumn harvest and is extensively utilized in Soviet agriculture.

## Veronal Buffer

Buffer is made from sodium barbitone and used to maintain the pH of fixative including glutaraldehyde used for the treatment of tissues for transmission or scanning electron microscopy.

## Vessel Element

Xylem molecule with specialized lignin thickening withinside the form of perforation plates. Vessels are seen best in angiosperms and are water-transporting in function.

## Vestigial

Descriptive of an organ that persists best as a vestige (i.e., in a reduced form) for the reason that organism wherein it occurs has so advanced that the organ is redundant. Human gill slits are vestigial organs.

## Villus

Finger-shaped protrusion from an epithelial surface. Examples are the intestinal villi, which greatly increase the absorption area of the intestine, and the chorionic villi, which ensure adequate contact between the placenta and the uterus. Note that villi are composed of many cells and should not be confused with micro-villas, which are subcellular and form a brush border in many absorptive cells.

## Vinblastine

Alkaloid from the Vinca plant inhibits spindle formation in dividing cells and is therefore used in cancer treatment. Vincristine is a similar but different compound.

## Vinculin

The fibrous protein is responsible for the attachment of actin filaments to the inner surface of the plasma membrane. It is distributed in patches known as adhesion plates and these are believed to be involved in intercellular adhesion.

## Viroid

The pathogen that causes plant diseases such as spindle tuber disease in potatoes and stalk growth in tomatoes. Viroids are single-stranded RNA molecules, although base-paired for much of their length, but unlike true viruses, are not associated with proteins.

## Virulence

The ability of a pathogenic organism to cause disease. Virulence is a function of invasiveness (the ability to colonize a host) and toxigenicity (the production of toxic substances such as exotoxins and endotoxins).

## Virulent Bacteriophage Phage

(e.g., *E. coli* phage T) which, upon infecting host bacteria, enters a viral growth cycle ending with the lysis of the bacterial cells and the release of progeny from the phage.

## Virus

Intracellular parasitic elements of plants, animals or bacterial cells, can only multiply in living cells. Viruses are likely "escaped" subcellular particles. They consist of one or more nucleic acid molecules within a protein envelope, or capsid made up of subunits or capsomeres.

Many viruses have an icosahedral (20-sided) structure, and some are also encased in an outer layer of membrane, which is a small part of the plasma membrane of the cell in which the virus replicates.

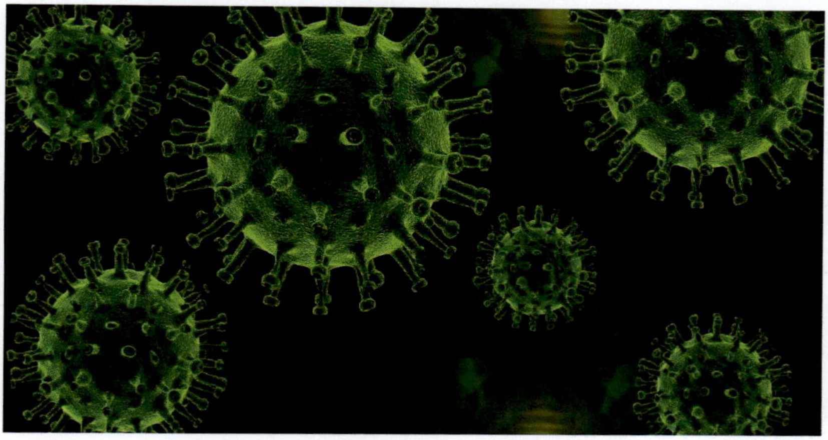

**Figure 69.** Image Showing A Virus.

*Source : Image By Pixabay*

## Viral Receptor

A specific protein, glycoprotein, or polysaccharide molecule on the surface of eukaryotic or bacterial cells is the specific site where viruses are absorbed.

## Vitamin A (Retinol)

A substance derived from plant carotenes; Nutritional deficiencies lead to night blindness and eventually stunted growth. It is stored in the liver and fish liver oil is a rich source of vitamins. Retinol aldehyde (retinen) is involved with opsin in the production of rhodopsin.

## Vitamin B Complex

Group of water-soluble vitamins that act as coenzymes or coenzyme precursors. Included in the grouping are 8-1 (thiamine), 8-2 (nicotinic acid and riboflavin), and 8-6 (pantothenic acid and pyridoxine), and 8-12 (cyanocobalamin and folic acid).

## Vitamin D

Fat-soluble steroid vitamins including D2 (calciferol) and D3, Effects of vitamin D include increased absorption of phosphate and calcium from the gut, and deficiency causes bone calcification abnormalities, including rickets. Vitamin precursors are converted to the active form in animal tissues in the presence of ultraviolet light. The plant precursors of D2 and D3 are ergosterol and 7-dehydrocholesterol.

## Vitamin D-Resistant Rickets

A sex-linked dominant genetic disease in humans with impaired renal phosphorus transport. The disease resembles rickets due to vitamin D deficiency but persists in the presence of adequate dietary vitamin D.

## Vitamin E (Tocopherol)

The group of fat-soluble vitamins consists of a-, b- and tocopherol and related molecules. The vitamin complex is found in some vegetables and particularly in wheat germ and is believed to act through membrane stabilization. A deficiency leads to reproductive disorders and infertility.

## Vitellogenin

Lipo-phosphoprotein: Precursor of the two main proteins lipo-vitelin and Phospho-vitin in egg yolk. In amphibians, vitellogenin (Mr 2,000,000) is synthesized in the liver of vitellogen females, secreted into the serum, transported to the ovaries, and included into growing oocytes by selective pinocytosis in small intracellular vesicles.

Inside the oocyte, the vesicles coalesce, vitellogenin is converted to lipo-vitelin and phosphor-vitin, and yolk proteins are organized to form yolk platelets.

## Vitelogenin Genes

Genes encoding Vitelogenin. Three of these genes exist in Drosophila, each with only one or two introns, while there are at least four distinct genes in *Xenopus laevis*. Two of the Xenopus A1 and A2 genes have been studied extensively. They are 21 and 16 kilobases long, respectively, both have 33 introns that interrupt the coding sequence, and both produce messenger RNA about 6.3 kilobases in length.

## Viviparous

Description of an organism capable of reproduction by retaining the embryo within the mother, who nourishes the embryo through placental connections. All placental mammals are viviparous, as are some snakes, lizards, sharks, and invertebrates.

## Voltinism

Variation in the presence or absence of disease (temporary cessation of growth) is evident as polymorphism in phyla of certain insects such as the silk moth *Bombyx mori*.

## Volutin

Cytoplasmic granules of polymerized inorganic phosphate appear in the cytoplasm of prokaryotic and eukaryotic microorganisms during nutrient starvation.

## Von Gierke Disease

Autosomal recessive disease in humans presents as defective storage of glycogen because of malformation of glucose-6-phosphatase. It has an incidence of one in 100,000.

# W

## Waardenburg Syndrome

An inherited disorder in humans with a frequency of about 1 in 42,000. It is inherited in an autosomal dominant manner and is characterized by severe deafness and other symptoms.

## Wallace Effect

Hypothesis suggested by Alfred Russell Wallace that populations of organisms tend to become reproductively isolated at the species level.

## Waltzer

A mutant strain of laboratory mice carrying a neurological defect. Homozygous animals tend to walk in circles and shake their heads.

## Warfarin

A compound used as a rat poison to treat venous thrombosisdue to its antagonism to blood coagulation. It is also a teratogen (can cause malformation) in humans emborys when taken during the first trimester of pregnancy.

## Waring

Blender homogenizer consists of rotating steel blades in a stainless-steel bowl; widely used to homogenize tissue samples and separate conjugated bacteria.

## Watson-Crick Model

Double Helix model for the structure of DNA proposed by J. Watson and F. Crick in 1953. The model proposed that two polynucleotide chains are linked by specific hydrogen bonds between complementary base pairs, with the sugar-phosphate backbones forming a double helix, thus providing a rigid backbone.

Although alternative structures for DNA were later discovered (e.g., Z-form left-handed). The model has proved correct and has provided tremendous encouragement and insight into the field of molecular genetics.

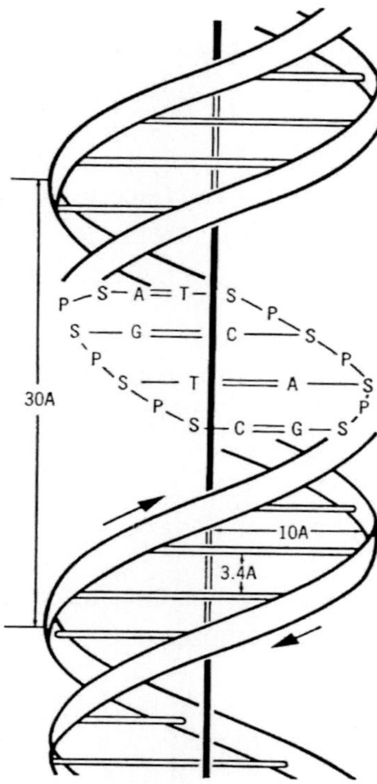


**Figure 70.** Double Helix DNA.

*Source : Image By Flickr*

## Waxes

Esters of long-chain fatty acids and alcohols are commonly found as protective outer coatings on plants and animals. For example, many leaves have a waxy cuticle, and bees use beeswax to form a honeycomb.

## Weissmann

Theorem. It is now generally accepted that characteristics acquired over the course of an organism's life are not inherited and that only genetic changes in germline cells (in animals) are passed on to offspring.

## Western Blotting

Protein identification method in complex mixtures. Proteins are separated by electrophoresis across a gel slab and then transferred to a nitrocellulose sheet by electrophoresis at 90 degrees to the gel surface. Proteins bound to the nitrocellulose membrane can then be identified by reaction with specific reagents such as radioactively or fluorescently labeled antibodies.

## Wheat Germ Lectin

A substance that specifically binds to N-acetylglucosamine and is used in cell biology to release these molecules from membranes.

## White Locus

Complex locus in the *Drosophila* genome affecting eye color. Many mutants have been identified at this locus, including some that contain regulatory rather than structural sequences.

## Whole-Arm Fusion

Fusion of the long arm of two chromosomes after breaking and loss of the small arms of two acrocentric chromosomes.

## Wild Type

A standard form found in wild or natural populations. The term can be used to refer to an individual's genotype or phenotype, or to a specific gene.

## Wild-Type Gene

Gene, which is the most common of all alleles at that locus and is therefore considered normal. Wild-type alleles are often represented as + and an organism homozygous for that gene would be represented as ++.

## Wilson's Disease

An autosomal recessive human disorder in which copper excretion is impaired and copper accumulation causes liver and brain damage.

## Wobble Hypothesis

Proposal by F. Crick to explain the fact that more than one triplet of codons can be recognized by a tRNA molecule carrying a single anticodon triplet. Degeneracy in the code means that in the mRNA some amino acids are specified by different codons but are recognized by the same aminoacyl-tRNA.

Almost all of the variation is in the last of the three letters in the codon, and the hypothesis proposes that jiggling at the anticodon-to-codon junction allows for some third-letter variation.

## W, Z Chromosomes

Specification of sex chromosomes in some organisms (e.g., silk moths) where the female is the heterogametic sex. Thus, the male is ZZ and the female is WZ, and the W chromosome is determinative of the female.

# X

## Xanthinuria

Human disease in which purine xanthine remains in the urine instead of being completely converted to uric acid.

## Xanthophylls

Class of oxygenated hydrocarbon compounds structurally related to carotene. Xanthophylls are yellow pigments, some of which function as photosynthetic auxiliary pigments in plants. In brown algae, however, the xanthophylls fucoxanthin and peridinin are the most important light-absorbing pigments.

## Xenogeneic Transplantation

Transplantation of tissue between organisms of different species, or the transfer of organelles from one cell of one species to another.

## *Xenopus laevis*

African clawed toad, an anuran clawed amphibian widely used in experimental embryology, gene expression control studies, and developmental immunology, where the existence of a free-living embryo (tadpole) and strains were histocompatibility inbreeds of *Xenopus laevis* particularly useful.

## Xeroderma Pigmentosum

Genetic metabolic disorder in humans in which DNA repair mechanisms are defective. One consequence is a high incidence of skin cancer due to the inability to repair damage caused by ultraviolet radiation from sunlight.

## Inactivation X

Repression by condensation of the transcriptional activity of one of the two copies of chromosome X in all female mammalian somatic cells. One of the two X chromosomes is randomly inactivated early in embryonic life as a mechanism of dosing offset.

The condensed chromosome is visible in the nuclei of somatic cells such as the Barr body, although some genes at the distal end of the short arm of the X chromosome can escape inactivation.

The coat color of the tortoiseshell cat results from the mosaic distribution of epidermal cell clones in which one of the X chromosomes is inactive and the pigment alleles are carried on that chromosome.

## XO

A situation in a heterogametic organism in which there is only a single X chromosome instead of an XX or XY. Individuals with turner syndrome are XO.

## X-Radiation

X-rays are produced when metal surfaces are bombarded with high-speed electrons. X-rays are ionizing radiation and have wavelengths between ultraviolet and gamma rays.

## X-Ray Crystallography

A technique that makes a specific contribution to molecular biology as it is the only method capable of revealing the three-dimensional position of atoms relative to each other in macromolecules such as proteins and nucleic acids.

The technique is based on the fact that X-rays have a wavelength of 0.154 nm, which is similar to intermolecular distances. X-rays pass through a crystal of the protein or other macromolecule, and those diffracted (scattered) by the crystal's atoms are detected on photographic film.

## Xylem

A complex cell tissue, consisting of different cell types, forms the most important water-conducting system in plants. The primary xylem forms first and develops from the primary meristem. It consists of protoxylem and metaxylem, with the former differentiating while the tissue is still expanding.

In plants with secondary thickening, the primary xylem is followed by the secondary xylem, which forms from the vascular cambium. The most characteristic cellular components are the trachear elements, which conduct water and also provide some mechanical support.

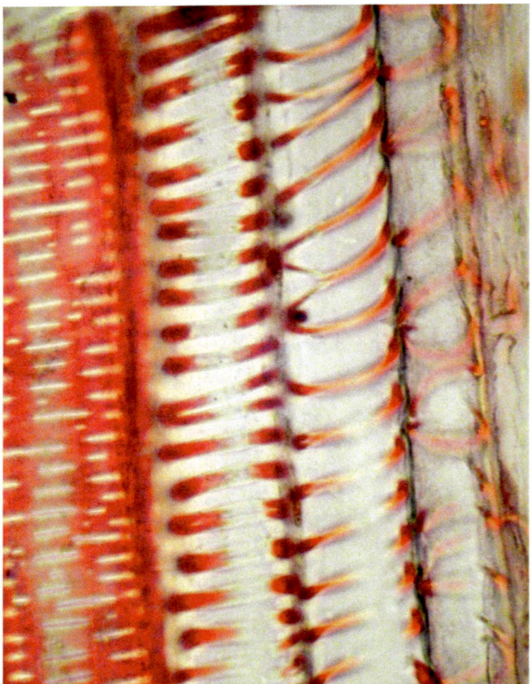

**Figure 71.** Longitudinal Section Of Xylem Vessels From Stem Of Sunflower (*Helianthus annuus*).

*Source : Image By Flickr*

# Y

## Yeast

Eukaryotic fungal cells are widely used and studied in cell biology and genetics. *Saccharomyces cerevisiae* is the budding yeast used in baking and brewing, and *Schizosaccharomyces pombe* is a fission yeast. Yeast cells grow rapidly in culture and have been used extensively to study the cell cycle.

Yeast small mutants, which have mitochondrial genetic mutations, are also being extensively studied, and yeast cells are now being used as expression systems for other eukaryotic gene sequences in recombinant DNA technology.

## Yeast Vectors

Yeast cells are utilized as carriers of plasmids, in which novel genes can be replicated, or transformed cells used to generate the products of integrated novel genes.

## Y Chromosomes

Chromosomes found in Branes, heterogametic sex. In most mammals, the male carries XY chromosomes and the female carries XX. The Y chromosome is a male determinant, as evidenced by the female phenotype of XO individuals Tturner's syndrome). It is a small heterochromatic chromosome with few or no structural genes on it.

## Y Linkage

Sexual linkage in which the genes are on the Y chromosome and are therefore only expressed in males (Dutch inheritance). In mammals, the Y chromosome carries few or no genes, but in some organisms (e.g., medaka fish) carries itsgenes on theY chromosome.

## Y -Suppressed Lethal Gene

Sex-linked recessive lethal gene allowing survival of XY males but causing the death of XO individuals in *Drosophila melanogaster*.

# Z

## Z Chromosome

Sex chromosomes are found in animal species where the female is heterogametic and therefore is WZ; Males are ZZ.

## Z DNA

Form of DNA in which the helix is twisted to the left instead of to the right. Z-DNA occurs naturally in short stretches of DNA.

## Zona Pellucida

A thin, noncellular covering secreted by follicular cells around the mammalian oocyte and persisting until the blastocyst stage just before implantation.

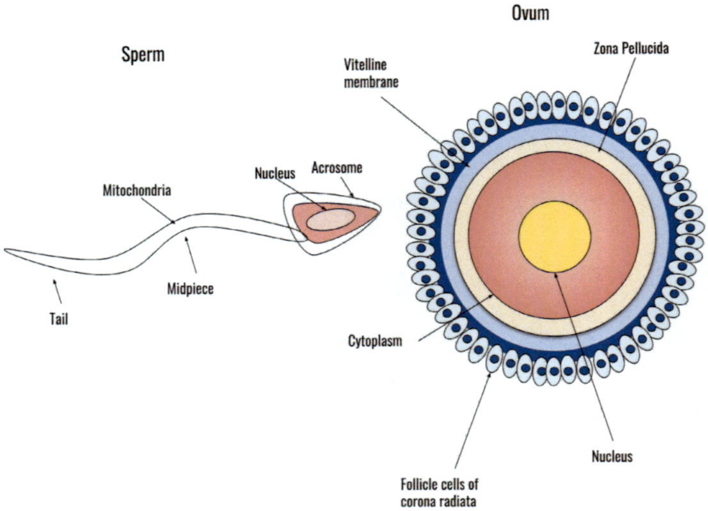


**Figure 72.** Sperm Entering The Ovum Using The Acrosome Which Breaks Down The Zona Pellucida With Enzymes.

*Source : Image By Wikimedia Commons*

## Zone Electrophoresis

Type of electrophoresis in which mobile ions or proteins in a buffer are displaced by a support medium such as paper, cellulose acetate, starch gel, silica gel, polyurethane foam, or acrylamide polymers. The sample is applied to the support in a narrow band approximately midway between the buffer reservoirs that house the electrodes.

During electrophoresis, each component migrates toward the cathode or anode at a characteristic rate, depending on the charge. With this method, the charged species are separated into discrete zones. The carrier can then be divided into sections and the individual components eluted separately.

## Zoospore

Flagellate spore is produced from a zoosporangium of many algae and some fungi. It is often, but not always, functional as a gamete.

## Zygotic Induction

Induction of a temperate phage present as a prophage on the chromosome of a donor bacterium into lytic growth when the prophage passes into a recipient cell during bacterial conjugation. Induction occurs because the recipient cell, which is not lysogenic, does not contain a phage repressor protein.

## Zymogen

The inactive form of a protein is usually a proteolytic enzyme. Zymogens are normally activated by post-translocation modification, often by cleavage and cleavage of part of the original polypeptide chain. Zymogens are often packaged into the cell, particularly the secretory cells of the pancreas, for example. These packages are called zymogen granules.

## Zymogen Granules

Used to have a specific meaning in relation to zymogen, but is now used as an ultrastructural term for any membrane-bound cytoplasmic storage granule.

## Zymogram

Electrophoresis gels to reveal and compare protein polymorphisms of isoenzymes such as lactate dehydrogenase subunits.

# REFERENCES


1. Brotherton, G., 2017. Key Concept: Equality. *Education Studies*, [online] pp.242-245. Available at: <https://www.researchgate.net/publication/329714117_Key_Concept_Equality_The_Key_Concepts> [Accessed 17 May 2022].
2. Burton, P., Tobin, M. and Hopper, J., 2005. Key concepts in genetic epidemiology. *The Lancet*, [online] 366(9489), pp.941-951. Available at: <https://www.researchgate.net/publication/7608812_Key_concepts_in_genetic_epidemiology> [Accessed 17 May 2022].
3. Panoutsopoulou, K. and Wheeler, E., 2018. Key Concepts in Genetic Epidemiology. *Methods in Molecular Biology*, [online] pp.7-24. Available at: <https://www.researchgate.net/publication/325598343_Key_Concepts_in_Genetic_Epidemiology> [Accessed 17 May 2022].
4. Patomäki, H., 2022. Key concepts. *The Three Fields of Global Political Economy*, [online] pp.7-19. Available at: <https://www.researchgate.net/publication/357897990_Key_concepts> [Accessed 17 May 2022].
5. Peim, N., 2017. Key Concept: Ontology. *Education Studies*, [online] pp.128-132. Available at: <https://www.researchgate.net/publication/329704693_Key_Concept_Ontology_The_Key_Concepts> [Accessed 17 May 2022].
6. Stein, H., 2021. Key Concepts. *Listening Deeply*, [online] pp.25-58. Available at: <https://www.researchgate.net/publication/355003983_Key_Concepts> [Accessed 17 May 2022].
7. Trotman, D., 2017. Key Concept: Creativity. *Education Studies*, [online] pp.35-39. Available at: <https://www.researchgate.net/publication/329704264_Key_Concept_Creativity_The_Key_Concepts> [Accessed 17 May 2022].

8. Trotman, D., 2017. Key Concept: Creativity. *Education Studies*, [online] pp.35-39. Available at: <https://www.researchgate.net/publication/329704264_Key_Concept_Creativity_The_Key_Concepts> [Accessed 17 May 2022].
9. Whisker, D., 2017. Key Concept: Hegemony. *Education Studies*, [online] pp.246-250. Available at: <https://www.researchgate.net/publication/329714017_Key_Concept_Hegemony_The_Key_Concepts> [Accessed 17 May 2022].
10. Whisker, D., 2017. Key Concept: Society. *Education Studies*, [online] pp.292-296. Available at: <https://www.researchgate.net/publication/329713807_Key_Concept_Society_The_Key_Concepts> [Accessed 17 May 2022].
11. Bobrow, M., 1998. A Dictionary of Genetics. *Journal of Medical Genetics*, [online] 35(2), pp.176-176. Available at: <https://www.researchgate.net/publication/24910726_A_Dictionary_of_Genetics> [Accessed 17 May 2022].
12. Boyd, P., 1987. A dictionary of genetic engineering. *FEBS Letters*, [online] 214(1), pp.201-201. Available at: <https://www.researchgate.net/publication/256172436_A_dictionary_of_genetic_engineering> [Accessed 17 May 2022].
13. Cassiman,, J., 2003. A Dictionary of Genetics. *European Journal of Human Genetics*, [online] 11(1), pp.105-105. Available at: <https://www.researchgate.net/publication/260177856_A_Dictionary_of_Genetics> [Accessed 17 May 2022].
14. Dodson, E. and Knight, R., 1950. Dictionary of Genetics. *American Midland Naturalist*, [online] 44(1), p.254. Available at: <https://www.researchgate.net/publication/275679200_Dictionary_of_Genetics> [Accessed 17 May 2022].
15. Harper, P., 1987. A Dictionary of Genetics. *Journal of Medical Genetics*, [online] 24(3), pp.190-190. Available at: <https://www.researchgate.net/publication/24909599_A_Dictionary_of_Genetics> [Accessed 17 May 2022].
16. Heslot, H., 1986. A dictionary of genetic engineering. *Trends in Biotechnology*, [online] 4(2), p.52. Available at: <https://www.researchgate.net/publication/277445338_A_dictionary_of_genetic_engineering> [Accessed 17 May 2022].

17. Hoy, M., King, R. and Stansfield, W., 1997. A Dictionary of Genetics. *The Florida Entomologist*, [online] 80(3), p.413. Available at: <https://www.researchgate.net/publication/275704943_A_Dictionary_of_Genetics> [Accessed 17 May 2022].

18. HU, Ph.D, J., 2000. A DICTIONARY OF GENETICS, FIFTH EDITION. *In vitro Cellular & Developmental Biology - Animal*, [online] 36(4), p.277. Available at: <https://www.researchgate.net/publication/257891411_A_dictionary_of_genetics> [Accessed 17 May 2022].

19. McGeoch, D., 1986. A dictionary of genetic engineering. *Endeavour*, [online] 10(1), p.52. Available at: <https://www.researchgate.net/publication/277518943_A_dictionary_of_genetic_engineering> [Accessed 17 May 2022].

20. Reardon, W., 1991. A Dictionary of Genetics. *Journal of Medical Genetics*, [online] 28(12), pp.888-888. Available at: <https://www.researchgate.net/publication/24883070_A_Dictionary_of_Genetics> [Accessed 17 May 2022].

21. Barrell, B., 1988. Gnomic: A dictionary of genetic codes. *Trends in Genetics*, [online] 4(10), p.296. Available at: <https://www.researchgate.net/publication/247106352_Gnomic_A_dictionary_of_genetic_codes> [Accessed 17 May 2022].

22. Brennan, J., 1969. Genetics A Dictionary of Genetics R. C. King. *BioScience*, [online] 19(6), pp.569-569. Available at: <https://www.researchgate.net/publication/270248160_Genetics_A_Dictionary_of_Genetics_R_C_King> [Accessed 17 May 2022].

23. Glass, B., 1972. <i>A Dictionary of Genetics.</i>Robert C. King. *The Quarterly Review of Biology*, [online] 47(4), pp.454-454. Available at: <https://www.researchgate.net/publication/247542746_Book_ReviewA_Dictionary_of_Genetics_Robert_C_King> [Accessed 17 May 2022].

24. HOLLANDER, W., 1961. A DICTIONARY OF GENETICS AND CYTOGENETICS. *Journal of Heredity*, [online] 52(6), pp.259-260. Available at: <https://www.researchgate.net/publication/299347359_A_dictionary_of_genetics_and_cytogenetics> [Accessed 17 May 2022].

25. King, R. and Stansfield, W., 2003. Book review: A dictionary of genetics, 6th edition. *Genome*, [online] 46(1), p.176. Available at: <https://www.researchgate.net/publication/237185734_Book_

review_A_dictionary_of_genetics_6th_edition> [Accessed 17 May 2022].

26. Maclean, N., 1987. Macmillan Dictionary of Genetics and Cell Biology. [online] Available at: <https://www.researchgate.net/publication/316267465_Macmillan_Dictionary_of_Genetics_and_Cell_Biology?_sg=5GLMOfL3cYskyuRlmU8bBwu8sbOLGLaWjMJ-z07Hcd5CYXXdzXm4sIx8CnUlhW-WXugz5ivX-hYAKvg> [Accessed 17 May 2022].
27. Matsudaira, T. and Hagiwara, M., 2004. Catchcopy Creation Support System Using Electronic Dictionary and Genetic Programming. *IEEJ Transactions on Electronics, Information and Systems*, [online] 124(1), pp.164-169. Available at: <https://www.researchgate.net/publication/315628533_Catchcopy_Creation_Support_System_Using_Electronic_Dictionary_and_Genetic_Programming> [Accessed 17 May 2022].
28. Matsudaira, T. and Hagiwara, M., 2004. Catchcopy Creation Support System Using Electronic Dictionary and Genetic Programming. *IEEJ Transactions on Electronics, Information and Systems*, [online] 124(1), pp.164-169. Available at: <https://www.researchgate.net/publication/315628533_Catchcopy_Creation_Support_System_Using_Electronic_Dictionary_and_Genetic_Programming> [Accessed 17 May 2022].
29. McGeoch, D., 1987. Gnomic: A dictionary of genetic codes. *Trends in Biochemical Sciences*, [online] 12, p.165. Available at: <https://www.researchgate.net/publication/247249127_Gnomic_A_dictionary_of_genetic_codes> [Accessed 17 May 2022].
30. Sorsby, A., 1969. Books and Monographs: A Dictionary of Genetics. *Journal of Medical Genetics*, [online] 6(2), pp.230-230. Available at: <https://www.researchgate.net/publication/25258918_Books_and_Monographs_A_Dictionary_of_Genetics> [Accessed 17 May 2022].
31. Temple, K. and Malcolm, S., 1987. A Dictionary of Genetic Engineering. *Journal of Medical Genetics*, [online] 24(11), pp.717-717. Available at: <https://www.researchgate.net/publication/24909945_A_Dictionary_of_Genetic_Engineering> [Accessed 17 May 2022].
32. Brand, N., 2004. Encyclopedic Dictionary of Genetics, Genomics and Proteomics, 2nd Edition. By George P. Rédei. *PROTEOMICS*,

[online] 4(2), pp.545-545. Available at: <https://www.researchgate.net/publication/246902865_Encyclopedic_Dictionary_of_Genetics_Genomics_and_Proteomics_2nd_Edition_By_George_P_Redei> [Accessed 17 May 2022].

33. Colin, E., 1949. <i>Dictionary of Genetics: Including Terms Used in Cytology, Animal Breeding and Evolution</i> . R. L. Knight. Waltham, Mass.: Chronica Botanica; New York: Stechert-Hafner, 1948. Pp. x + 183. $4.50. *Science*, [online] 109(2834), pp.417-418. Available at: <https://www.researchgate.net/publication/252305335_Book_Reviews_Dictionary_of_Genetics_Including_Terms_Used_in_Cytology_Animal_Breeding_and_Evolution> [Accessed 17 May 2022].

34. Cuéllar Lázaro, C., Mallo Lapuerta, A. and Córdova Martínez, A., 2014. Designing a multilingual dictionary of genetic terms (English, French, German and Spanish) for the European Portal Eurogene and the International Scientific Community. *Journal of Human Sport and Exercise*, [online] 9(4), pp.773-781. Available at: <https://www.researchgate.net/publication/277655950_Designing_a_multilingual_dictionary_of_genetic_terms_English_French_German_and_Spanish_for_the_European_Portal_Eurogene_and_the_International_Scientific_Community> [Accessed 17 May 2022].

35. Elias, J., 1998. <i>A Dictionary of Genetics.</i>Robert C. King , William D. Stanfield. *The Quarterly Review of Biology*, [online] 73(1), pp.71-72. Available at: <https://www.researchgate.net/publication/247551922_Book_ReviewA_Dictionary_of_Genetics_Robert_C_King_William_D_Stanfield> [Accessed 17 May 2022].

36. Falkinham, J., 1999. Book review. A Dictionary of Genetics, 5th edition. Robert C King, William D Stansfield. *Journal of Heredity*, [online] 90(4), pp.504-504. Available at: <https://www.researchgate.net/publication/31284943_Book_review_A_Dictionary_of_Genetics_5th_edition_Robert_C_King_William_D_Stansfield> [Accessed 17 May 2022].

37. Gareth Johnson, 2004. Encyclopedic Dictionary of Genetics, Genomics and Proteomics (2nd ed.). *Reference Reviews*, [online] 18(6), pp.43-44. Available at: <https://www.researchgate.net/publication/280185459_Encyclopedic_Dictionary_of_Genetics_Genomics_and_Proteomics_2nd_ed2004333George_P_Redei_Encyclopedic_Dictionary_of_Genetics_Genomics_and_Proteomics_2nd_ed_

Hoboken_NJ_Wiley-Liss_2003_xi1379_pp_ISBN_0_4> [Accessed 17 May 2022].

38. Glass, B., 1985. <i>A Dictionary of Genetics.</i>Robert C. King , William D. Stansfield. *The Quarterly Review of Biology*, [online] 60(3), pp.348-349. Available at: <https://www.researchgate.net/publication/247547694_Book_ReviewA_Dictionary_of_Genetics_Robert_C_King_William_D_Stansfield> [Accessed 17 May 2022].

39. Reeve, E., 1991. A Dictionary of Genetics, 4th Edition. By Robert C. King and William D. Stansfield. Oxford University Press. 1990. 406 pages. Paperback £16.00, ISBN 0 19 506371 6. *Genetical Research*, [online] 58(1), pp.92-93. Available at: <https://www.researchgate.net/publication/240214380_A_Dictionary_of_Genetics_4th_Edition_By_Robert_C_King_and_William_D_Stansfield> [Accessed 17 May 2022].

40. *The American Biology Teacher*, 1987. A Dictionary of Genetic Engineering Stephen G. Oliver John M. Ward. [online] 49(6), pp.389-389. Available at: <https://www.researchgate.net/publication/314827729_Review_A_Dictionary_of_Genetic_Engineering_by_Stephen_G_Oliver_John_M_Ward> [Accessed 17 May 2022].

41. Truman, D., 1985. A Dictionary of Genetics. Third edition. By R. C. King and W. D. Stansfield. Oxford University Press. 1985.480 pages. £15.00 paperback, £25.00 hardback. ISBN 0 19 503495 3; ISBN 0 19 503494 3. *Genetical Research*, [online] 46(2), pp.238-239. Available at: <https://www.researchgate.net/publication/247376998_A_Dictionary_of_Genetics_Third_edition_By_R_C_King_and_W_D_Stansfield> [Accessed 17 May 2022].

# INDEX

**S**



**T**



**U**



**V**



**W**



**X**



**Y**



**Z**

FSC
www.fsc.org
MIX
Papier | Fördert
gute Waldnutzung
FSC® C083411

Kolibri 360 GmbH
Ferdinand-Jühlke-Straße 7
99095 Erfurt, Deutschland
produktsicherheit@kolibri360.de